ENSEMBLE MODELING

STATISTICS: Textbooks and Monographs

A SERIES EDITED BY

D. B. OWEN, Coordinating Editor

Department of Statistics
Southern Methodist University
Dallas, Texas

Vol. 1: The Generalized Jacknife Statistic, *H. L. Gray and W. R. Schucany*
Vol. 2: Multivariate Analysis, *Anant M. Kshirsagar*
Vol. 3: Statistics and Society, *Walter T. Federer*
Vol. 4: Multivariate Analysis: A Selected and Abstracted Bibliography, 1957-1972, *Kocherlakota Subrahmaniam and Kathleen Subrahmaniam* (out of print)
Vol. 5: Design of Experiments: A Realistic Approach, *Virgil L. Anderson and Robert A. McLean*
Vol. 6: Statistical and Mathematical Aspects of Pollution Problems, *John W. Pratt*
Vol. 7: Introduction to Probability and Statistics (in two parts), Part I: Probability; Part II: Statistics, *Narayan C. Giri*
Vol. 8: Statistical Theory of the Analysis of Experimental Designs, *J. Ogawa*
Vol. 9: Statistical Techniques in Simulation (in two parts), *Jack P. C. Kleijnen*
Vol. 10: Data Quality Control and Editing, *Joseph I. Naus* (out of print)
Vol. 11: Cost of Living Index Numbers: Practice, Precision, and Theory, *Kali S. Banerjee*
Vol. 12: Weighing Designs: For Chemistry, Medicine, Economics, Operations Research, Statistics, *Kali S. Banerjee*
Vol. 13: The Search for Oil: Some Statistical Methods and Techniques, *edited by D. B. Owen*
Vol. 14: Sample Size Choice: Charts for Experiments with Linear Models, *Robert E. Odeh and Martin Fox*
Vol. 15: Statistical Methods for Engineers and Scientists, *Robert M. Bethea, Benjamin S. Duran, and Thomas L. Boullion*
Vol. 16: Statistical Quality Control Methods, *Irving W. Burr*
Vol. 17: On the History of Statistics and Probability, *edited by D. B. Owen*
Vol. 18: Econometrics, *Peter Schmidt*
Vol. 19: Sufficient Statistics: Selected Contributions, *Vasant S. Huzurbazar (edited by Anant M. Kshirsagar)*
Vol. 20: Handbook of Statistical Distributions, *Jagdish K. Patel, C. H. Kapadia, and D. B. Owen*
Vol. 21: Case Studies in Sample Design, *A. C. Rosander*
Vol. 22: Pocket Book of Statistical Tables, *compiled by R. E. Odeh, D. B. Owen, Z. W. Birnbaum, and L. Fisher*
Vol. 23: The Information in Contingency Tables, *D. V. Gokhale and Solomon Kullback*
Vol. 24: Statistical Analysis of Reliability and Life-Testing Models: Theory and Methods, *Lee J. Bain*
Vol. 25: Elementary Statistical Quality Control, *Irving W. Burr*
Vol. 26: An Introduction to Probability and Statistics Using BASIC, *Richard A. Groeneveld*
Vol. 27: Basic Applied Statistics, *B. L. Raktoe and J. J. Hubert*
Vol. 28: A Primer in Probability, *Kathleen Subrahmaniam*
Vol. 29: Random Processes: A First Look, *R. Syski*

Vol. 30: Regression Methods: A Tool for Data Analysis, *Rudolf J. Freund and Paul D. Minton*
Vol. 31: Randomization Tests, *Eugene S. Edgington*
Vol. 32: Tables for Normal Tolerance Limits, Sampling Plans, and Screening, *Robert E. Odeh and D. B. Owen*
Vol. 33: Statistical Computing, *William J. Kennedy, Jr. and James E. Gentle*
Vol. 34: Regression Analysis and Its Application: A Data-Oriented Approach, *Richard F. Gunst and Robert L. Mason*
Vol. 35: Scientific Strategies to Save Your Life, *I. D. J. Bross*
Vol. 36: Statistics in the Pharmaceutical Industry, *edited by C. Ralph Buncher and Jia-Yeong Tsay*
Vol. 37: Sampling from a Finite Population, *J. Hajek*
Vol. 38: Statistical Modeling Techniques, *S. S. Shapiro*
Vol. 39: Statistical Theory and Inference in Research, *T. A. Bancroft and C.-P. Han*
Vol. 40: Handbook of the Normal Distribution, *Jagdish K. Patel and Campbell B. Read*
Vol. 41: Recent Advances in Regression Methods, *Hrishikesh D. Vinod and Aman Ullah*
Vol. 42: Acceptance Sampling in Quality Control, *Edward G. Schilling*
Vol. 43: The Randomized Clinical Trial and Therapeutic Decisions, *edited by Niels Tygstrup, John M. Lachin, and Erik Juhl*
Vol. 44: Regression Analysis of Survival Data in Cancer Chemotherapy, *Walter H. Carter, Jr., Galen L. Wampler, and Donald M. Stablein*
Vol. 45: A Course in Linear Models, *Anant M. Kshirsagar*
Vol. 46: Clinical Trials: Issues and Approaches, *edited by Stanley H. Shapiro and Thomas H. Louis*
Vol. 47: Statistical Analysis of DNA Sequence Data, *edited by B. S. Weir*
Vol. 48: Nonlinear Regression Modeling: A Unified Practical Approach, *David A. Ratkowsky*
Vol. 49: Attribute Sampling Plans, Tables of Tests and Confidence Limits for Proportions, *Robert E. Odeh and D. B. Owen*
Vol. 50: Experimental Design, Statistical Models, and Genetic Statistics, *edited by Klaus Hinkelmann*
Vol. 51: Statistical Methods for Cancer Studies, *edited by Richard G. Cornell*
Vol. 52: Practical Statistical Sampling for Auditors, *Arthur J. Wilburn*
Vol. 53: Statistical Signal Processing, *edited by Edward J. Wegman and James G. Smith*
Vol. 54: Self-Organizing Methods in Modeling: GMDH Type Algorithms, *edited by Stanley J. Farlow*
Vol. 55: Applied Factorial and Fractional Designs, *Robert A. McLean and Virgil L. Anderson*
Vol. 56: Design of Experiments: Ranking and Selection, *edited by Thomas J. Santner and Ajit C. Tamhane*
Vol. 57: Statistical Methods for Engineers and Scientists. Second Edition, Revised and Expanded, *Robert M. Bethea, Benjamin S. Duran, and Thomas L. Boullion*
Vol. 58: Ensemble Modeling: Inference from Small-Scale Properties to Large-Scale Systems, *Alan E. Gelfand and Crayton C. Walker*

OTHER VOLUMES IN PREPARATION

ENSEMBLE MODELING

Inference from Small-Scale Properties to Large-Scale Systems

Alan E. Gelfand
Department of Statistics

Crayton C. Walker
Department of Information Management

The University of Connecticut
Storrs, Connecticut

MARCEL DEKKER, INC. New York and Basel

Library of Congress Cataloging in Publication Data

Gelfand, Alan E., [date]
Ensemble modeling.

(Statistics, textbooks and monographs ; v. 58)
Includes index.
1. System analysis. 2. Mathematical models.
I. Walker, Crayton C., [date]. II. Title.
III. Series.
QA402.G435 1984 519.5 84-12037
ISBN 0-8247-7180-X

MARCEL DEKKER, INC.
270 Madison Avenue, New York, New York 10016

Current printing (last digit):
10 9 8 7 6 5 4 3 2 1

PRINTED IN THE UNITED STATES OF AMERICA

To M. E. and H. R.

Preface

To set this book in perspective, and to suggest the proper audience for it, we note that a particular modeling scheme is followed in several specific areas. This puts it in general systems theory, if that term is understood as the theory of midlevel modeling. This definition is essentially that proposed in 1956 by Boulding*, who described general systems theory as the study of modeling conducted at levels of abstraction lower than that of pure mathematics, but higher than those used in specific disciplines. Several other threads of emphasis lace through the work: (1) The modeling approach taken juxtaposes stochastic and deterministic processes; (2) we explicitly introduce an observer's perspective into the modeling process by the use of populations of models (called "ensembles" in what follows); and (3) we maintain a rather sharp representative model versus theoretical model distinction, coming down on the side of theory by our insistence on simple models. Hence we address ourselves especially to

---

*K. Boulding, General systems theory--The skeleton of science. *Management Science, 2,* 197-208 (1956). Also found in J. Buckley (Ed.), *Modern Systems Research for the Behavioral Scientist*. Chicago: Aldine, 1968, pp. 3-10.

general systems theorists, modeling theorists, statisticians, and probabilists.

Since we work with switching nets, and do some combinatorics (on Boolean functions), readers with backgrounds in these areas might find something of interest --switching theorists primarily because we try to extend the interpretive reach that their models have.

This work is marked by our interests in the properties of formal models. Nevertheless, we devote a fair amount of effort to interpretation. Our persuasion is that the subject-matter specialist deserves at least some hints as to how the abstract systems examined here might be linked to the real world, however wide of the mark our efforts may turn out to be. We discuss an existing biological interpretation, and go on to suggest applications of ensemble-based network modeling in organizational and marketing contexts. In addition to these motives in presenting interpretive material, we have been impressed with the clarity that the effort of interpreting simple models seems to bring to the subject matter at hand, and for that reason recommend the exercise to substantive theorists.

The scope of our discussion is limited. We do not attempt to review all uses of network models. The bulk of the literature on graph theory, for example, is not of interest here. Our emphasis is on how locally specifiable properties of models influence overall model behavior. In this modeling the behavioral characteristics of the network elements themselves, Boolean transformations, are of focal importance.

Limited as our subject is, at present pieces of it appear in widely scattered sources. In this book we wanted to bring these pieces together, add to them where we could, but especially to provide a framework in which these separate contributions could be seen as more of an intellectual unity.

We have tried to make the mathematics used accessible to the nonspecialist. We introduce the mathematical section in such a way that only relatively basic tools in probability and mathematical statistics will suffice for understanding. Readers resolutely disinterested in mathematical details are encouraged to concentrate on the interpretive material. Such readers may find Chapters 1, 3, and 6 of most interest.

To summarize content by chapters: Chapter 1 states our general goals, then gives our view of modeling, pointing out that complex things, simple models, and ensembles go usefully together. In Chapter 2 we try to characterize more formally our ensemble approach by showing how it differs from conventional data-fitting approaches. We are not estimating parameter values in the traditional sense, but selecting plausible subsets of models. Chapter 3 introduces terminology and gives a rudimentary development of the models used, together with a preview of the application areas described later. Chapter 4 gives a technical development of the mathematics now available for deducing network behavior. Chapter 5 summarizes simulation results most pertinent for our discussion. Chapter 6 displays the potentially wide scope of a net ensemble modeling approach using examples in several substantive areas.

Finally, this manuscript is a testimonial to collaborative research work. We are two scientists with very different training who individually would have been limited in vision or in technical skill, but who have jointly flourished. Each of us made contributions in accordance with our expertise and in the end we hope the reader will find that a sound, coherent, provocative tract has emerged. Needless to say, we are solely responsible for any inaccuracies or errors.

We acknowledge the Office of Naval Research through Herbert Solomon at Stanford University, the Center for Real Estate and Urban Economic Studies at the University of Connecticut, and the University of Connecticut Research Foundation for partial support of this project.

A major portion of the typing and preparation was done by Katharine Holmes. Additional typing was done by Jeanne Young and Carolyn Knutsen. We are grateful for their able efforts.

Alan E. Gelfand
Crayton C. Walker

Contents

ENSEMBLE MODELING

1

Introduction

1.1 Main Questions

Our overall purpose is to examine several broad questions in general systems theory, and to suggest areas of application for the approach used. The main questions are: In systems where details matter, what does limited specification of those details mean for the predictability of overall system dynamics? For example, do details of system structure (in a well-defined sense) matter? Is explanation of system dynamics decisively thwarted by lack of knowledge of underlying details? Could explanation perhaps be *aided* instead? If so, in what circumstances might that be? Can certain limitations on specification be usefully interpreted in the real world? Would *control* of system behavior make practical sense in such settings?

The success of statistical mechanics in, say, the study of gases suggests an answer to at least some of these questions. Predictability without using all details is, to a degree, possible. Explanation is also possible.

What we are doing here that is different is to show how a statistical approach to these questions can still be applied to systems in which their stochastic nature is less apparent than their determinism. In this book we are concerned with "nongaseous" complex systems, systems in which transactions among their many parts follow unchanging (but complicated and otherwise largely unknown) interaction paths. In addition to discussing an established application in biology, we intend to move this statistical method--an ensemble approach--into an important realm of internally organized and selective systems: the social realm.

Among the broad findings made using this approach is that some behavioral regularities are robust with respect to changing internal organizational details. Additionally, control methods and explanatory variables can be identified in situations where internal change, or an observer's ignorance of internal detail, may appear overwhelming.

More specifically, we first examine and extend an existing ground-breaking model of the genetic control system. It is argued (see, e.g., Kauffman, 1974) that the biochemical identity of a biological cell is specified by the sequential activity of the genes within it. These activities are determined by the genes themselves through the intricate interaction pattern by which they repress or "derepress" one another's productions. Approximating this genetic apparatus with a binary switching net model, and studying the properties of the model, one can directly give reasons why it is advantageous, in the evolutionary sense, for each gene to be affected by only a very limited number of other genes, why tissue types differentiate into no more than a half-dozen or so of other tissue types, why genetic activities are largely homeostatic, and why the number of tissue types increases as a fractional power of the number of genes in an organism.

In the managerial realm we argue that in some organizations, control mechanisms can usefully be seen as switching nets. So construed, we are able to present evidence that the productive routine of such organizations can be maintained in a wide variety of structural circumstances, provided that certain loosely specified constraints on other aspects of organizational detail obtain. We will suggest parallels between characteristics of our generic models (switching nets) and existing or potential real-world management principles. Those we examine are: span of control, the exception principle, the scalar principle of management, and consensus-level management. Using our modeling approach we are able to find relationships among and move toward quantification and theoretical clarification of these management principles. We point out, for example, how the general utility of the exception principle can be explained in this context. We examine the interesting question of how organizations built only by reference to small scale detail should differ from organizations constructed on a fully planned basis. We show how psychological constructs related to organizational climate might be handled in our scheme, and, in a frankly speculative effort, suggest the possible psychological effects of the managerial control styles with which we are dealing.

Moving our model networks into a marketing and advertising context where nets are seen as groups of interacting consumers, we show how changes in net characteristics can be interpreted as types of advertising message content. Those we define are: primary persuasion, imitative persuasion, and consensus persuasion. We then examine the theoretical effects that advertising campaigns making use of these different persuasive thrusts would have on (1) buyer group disposition toward an opinion target and on (2) those buyer groups' brand loyalty, as interpreted in our modeling scheme. Our model suggests, for example, that where relatively enduring enhancement of opinion is sought, an extended advertising campaign making use of consensus or imitative persuasion is desirable. Where a quick but possibly transient opinion boost would suffice, a short campaign making use of high intensity primary persuasion is indicated.

Since the network ensemble modeling approach we use may be unfamiliar, we begin with a general discussion of modeling as a way of introducing the specifics of our approach. The reader preferring to move directly into the model development may wish to proceed to Chapter 3. The balance of Chapter 1 considers the philosophy of modeling, while Chapter 2 clarifies the inferential framework of our modeling approach. In any event, the reader should feel free to read selectively in what follows.

1.2 Models

Dictionary definitions of "model" are surprisingly varied. The term can signify a small replica of something, a standard that should or may be followed, or something which is to be copied. Common usage allows the term to refer both to a copy of something and to something that is to be copied. Our focus will be on models in the first sense: as means by which things are duplicated in different forms.

It is significant that it is difficult to specify what a model is and what it is not. Part of the difficulty lies in the fact that a model is a model of something, and for some purpose. Burks (1975) points out that a model is a triadic relation involving that which is modeled, the model itself, and the purpose of the model. All three parts are needed for a complete specification. For example, is a 150-pound sack of sand a model? Not necessarily, of course, but it could be a model of a human being in a study of automobile dynamics. This is to say that a model isolated in a sense does not exist. It is just another thing: perhaps a collection of statements or equations.

An interesting illustration of this triadic relation in common use is provided by some riddles. In a riddle, a thing or a situation is given, usually in verbal form. The game can be to provide the unstated parts of the triad so as to make the thing become a model, and hence, to make sense. What is white and black, has one horn, and gives milk? What moves on four legs in the morning, two at noon, and three in the evening?*

Modeling is a very pervasive human activity. Any time we describe something, to ourselves or to someone else, we are modeling. Description provides a way in which the thing modeled can be brought into the social order in a controlled, accessible, and transferable manner. That is, a model of something, its description-for-some-purpose, provides for selective, focused perception of the thing in forms that allow symbolic manipulation, both public and private. In short, a model allows us to discuss the thing, with others, or with ourselves. Modeled, the thing can participate in a variety of public processes. In this form, our perceptions of it can be debated and thereby sharpened or broadened. Finally, in this form the thing can be understood. That description provides such benefits is obvious. That descriptions of things are possible is one of the true

---

*A milk truck and a man, respectively. The first is a children's riddle; the second is from Greek mythology. In these riddles the modeling purpose appears to be description: in the first case a simple visual description is intended, in the second, a less obvious temporal relationship. Clearly, an important part of the entertainment function of riddles lies in an apparent misspecification of the original by the stated model.

natural wonders.

Description involves abstracting, selecting features for emphasis from among the myriad of possibilities inherent in the thing itself. But abstractions, to be communicated, must be expressed in some way. Hence a model may be an artifact (a toy train, a set of equations on a chalkboard) or a natural object (a tree offered as a model of a river). In either case, whether constructed or found, the thing offered as a model serves as a model by virtue of its expression of some abstracted content. (For an enjoyable illustration of some geometrical similarities, see Stevens, 1974.) Models, then, have binary nature. They are both real things and abstractions.* Take the example of a tree offered as a model of a river. The abstractions being carried may be obscured by being left unremarked on. The model clearly has other features. Are the leaves of interest, the roughness of the bark, birds' nests? In the present example, it is surely a two-dimensional representation of the tree's branching pattern that is being pointed to. Moreover, it is almost certainly the case that what is common among the patterns of many trees is being remarked on as resembling what is common in many river systems' branching patterns. Some models' abstracted content is more implied than explicit, and

---

*This can be a matter of some importance in psychopathology. "The schizoid confusion of symbols with the thing symbolized . . ." (Murphy, 1967, p. 424) may be relevant in this discussion of models, in that models, while being representative, can also be used as symbols (which ordinarily need not decisively resemble the thing symbolized). The schizoid person, owing to difficulty in social functioning, may not perceive the purposive aspect of models. That is, he or she may fail to separate model and original because of an inability or refusal to deal with the other-person aspect of models: hence a person who speaks in riddles.

further, there may also exist the implication of an abstractive process that is statistical in nature.

1.3 Models Versus Theories

Modeling or theorizing: which are we doing? Both, it turns out, but perhaps more of the second than the first. This should be explained, since the terms "model" and "theory" are often used interchangeably (Simon and Newell, 1956).

Models of real things are abstractions, simplifications of reality. So are theories, in our view. In that both are representative simplifications, models and theories are similar. It is possible that a set of abstractions might be at the same time a (representative) model and a theory, depending on the purpose for which the set is being used at the given time. From our point of view, however, they are importantly different. What distinguishes a theory from a model is the analyst's purpose. Theories are meant to be understood: to explain something. Models are meant to be apprehended: to display something. What is to be displayed may be as routine as a simple physical appearance, or as complex as a intricate behavioral sequence. The best model, best in the sense of resembling the original, is an arbitrarily close duplicate of the original. That is, the "best" model is a set of abstractions elaborated to the point where no differences can be detected between the model and the original. But in being virtually indistinguishable from the original, it has at the same time contracted virtually all the complexity of the original as well. It has become useless for transmitting an explanation of the original, and therefore, it is now as bad a theory of the original as can be devised. To the extent that a model is only required to mimic, while a theory is additionally supposed to explain, to that extent they may decisively differ. In more picturesque language, a good model is like a fine portrait. A good theory is like a masterful caricature.

The difference these requirements make in the content of models and theories is, for our present purposes, largely a matter of the theory being an artfully reduced model. The analyst may first model descriptively in the hope that the model will include a theory that is easy to discern. That is, the analyst may add features to construct a descriptive model and subtract features from a good model to make a theory. We shall examine below whether or not this is the only way to construct theories.

In the world of practical affairs, a model is often all that is required. Decisions can be made if a situation can be conceptualized sufficiently so that prediction is possible. This conceptualization, or model, may contain features that are actually unnecessary. On the other hand, if the analyst wants to understand something, to know why it behaves the way it does, then the demand is not only for a set of abstractions that is sufficiently powerful to simulate the original's behavior, but in addition, for a set which is necessary in that behavior. To understand something, it is common to ask for those properties which must exist: for those without which the model would fail decisively to resemble the original. Thus a demand for explanation, in our view, carries with it a strong flavor of reduction. This flavor permeates science for good reason.

Since the reduction of a model is a step beyond what is required for many practical purposes, it may appear to be an excessive and pointless activity to people who are practically oriented. However, the need for knowledge concerning reduced models, for "basic" science, does appear in the world of work. It appears, for example, when the manager wants a more efficient way of doing something. To make a process more efficient, or cheaper, it helps to know what factors are indispensable and which are not. Theory is not irrelevant in practice.

Coombs et al. (1954, p. 137) also distinguish between a model and a theory:

> It might be well to draw clearly the distinction between a model and a theory. A model is not of itself a theory; it is only an available or possible or potential theory until a segment of the real world has been mapped into it. Then the model becomes a theory about the real world. As a theory, it can be accepted or rejected on the basis of how well it works. As a model, it can be right or wrong only on logical grounds. A model must satisfy only internal criteria; a theory must satisfy external criteria as well.

The discussion above can be more readily understood if it is known that the authors mean by "model" essentially a set of mathematical axioms and postulates which may be quite abstract and uninterpreted. Their "theory" comes out as the consequence of exercising the set of mathematical statements and linking these elaborations to the real world. Their model then, is intended to expose the logical foundations of the theory. More exactly, it is intended to expose a logical foundation, since there is no guarantee that any one foundation is unique. At any rate, once a foundational structure is achieved, it is easier for questions of independence of axioms, or relations between foundational models, and so on, to be answered. These are important questions, to be sure. However, it should be noted that a foundational model of the Coombs sort, while it may make certain aspects of a theory understandable, may not contribute any insight about the portion of the real world which is the subject of the theory. Axiom systems are notorious in this regard. For the simple reason that we believe that the term "model" should refer to abstractions that are genuinely representative, we believe that the term should not be used in contexts where representation is neither intended nor facilitated.

We emphatically disagree with the point of view which holds that if one can model a phenomenon, then one has a theory of that phenomenon. In the early days of computers, it was not uncommon to hear it claimed that a computer program simulating some behavior *is* a theory of that behavior. Admittedly, discussing something dimly understood, or otherwise modeling it, may well assist in advancing one's understanding of it. Discussing some phenomenon with a computer, in the sense of programming the computer to display the dynamics of the phenomenon, may be extraordinarily instructive, if for no other reason than that the computer use requires one's own ideas to be set forth in exquisite detail. However, even given that complex models can be relatively easily constructed by programming, it does not follow that understanding the resulting programs will necessarily be easy. The recent emphasis on "structured programming" in the computer field came about through recognition of the difficulties that await someone who tries to understand even a fair-sized program. More on this point below.

1.4 Description

It is clear that we agree with Caswell when he says: "In particular, I will distinguish between (i) models that are constructed primarily to provide accurate prediction of the behavior of a system, and (ii) models that, as scientific theories, are attempts to gain insight into how the system operates" (Caswell, 1976,

p. 317). That is, as we have already argued, models may serve two distinct purposes, description (which we take to include prediction since that is just the description of a state of affairs which is to be) and explanation. Further, since "theory" in both common and in scientific usage implies an attempt at explanation, a theory is a model that not only describes, but explains.

The discussion above has suggested that theories are more than just descriptive models. This strict "add-on" view of the nature of theories, that is, seeing them as constructed by acting on full-blown representational models, may not be what can, or should, occur typically. As we shall discuss below, and have already implied, there very likely exists some trade-off between description and explanation. The supremely representative model is likely to be difficult to understand. The transparently understandable model may well sacrifice representation. "We intuitively feel that with increasing complexity a model will become more and more descriptive rather than explanatory . . ." (Schnakenberg, 1977, p. 2). To the extent that one cannot have both prediction and explanation in a given modeling context, theories will not simply add explanatory content to descriptive content. It is this descriptive-explanatory trade-off that in the end makes models and theories disparate.

1.5 Explanation

Our use of "description" should be clear by this time. It means essentially what others mean by models' "fitting the data." We now take up the question of how we view explanation.

The first point to be made is that the modeling context, which we have referred to as the purpose of the model, is important in explanation. In particular, we can set aside, to use Simon's term (1973, p. 24), the Laplacian program of explaining a phenomenon by presenting the "fundamental" microscopic equations of all the physical particles involved. Indeed, we can set this program of "ultimate" explanation aside as metaphysical, since science is a matter of theory, theories are models, and models exist in specific contexts which decisively involve purpose and human utility. That is, we accept the sociological character of science as basic to its nature. Moreover, we are convinced that this aspect of science need not be apologized for. As modeling is carried out with respect to a particular context, since explanation makes use of models, it too is context dependent.

We now consider sufficient conditions for a model's fitting the data. As an example, consider an intelligent individual, almost totally unfamiliar with masonry technique, who has asked himself: What must I do to build a wall with these bricks? If our protomason intends literally to answer the question (and not to simply pile bricks upon bricks in an unreflective way), he will need (1) a wall, or at least a conception of a wall, and (2) a model of how the wall is built. That is, if he is actually to articulate the answer, he needs (1) a source of real-world data to compare with (2) the results predicted by his model. He will, of course, try various models, discarding those that do not "fit" the real world in the sense of producing a wall when their prescriptions are followed. In fact, he may first set for himself the intermediate goal of answering the question: How can I support one brick with other bricks? Let us assume that he sets out to answer this question by taking bricks A and B, and setting brick C on top of them in various orientations. He finds that some arrangements support brick C, others do not.

Finally, he hits on this model: A and B set a distance apart that is 1 inch less than C's length will support C if it is placed equally on both A and B. This model appears to work. C is, in fact, supported in this condition. Further, it appears that this model, elevated to serve as a construction technique, gives a pleasing regularity of appearance and allows itself to be repeated both horizontally and vertically. Of course, as far as his data go, such an elevation of the specific model makes at least two inductive leaps. His model makes use of bricks A, B, and C. Whether *any* three bricks will fit the model (i.e., whether any two bricks a distance C less 1 inch . . . will support the third) depends on how much the bricks in the collection he has resemble the three bricks he chose at the start. And whether the n-th row of bricks will be supported depends on factors he has not evaluated to this point. But supposing conditions favorable, he has developed a procedure that suffices for wall building in a rudimentary way. We can say that he has a sufficient model.

It is clear that a sufficient model "consists of" information sufficient to provide a resemblance that meets some criterion of fit or use. Here, the resemblance to be judged would be that between what is built and what the builder or others hold to be walls.

Explanation is more complicated. An explanation is an answer to a "why" question. Consider the following situation. An adult driving an automobile is asked by a child: "Why does the car go?" Since a discussion is being initiated, a model is required. Let us assume that the adult decides that the modeling context is such as to make appropriate a model emphasizing present circumstances: an automobile moving under power on a level road, and readily apparent operator controls:

accelerator, brake pedal, shift lever, and so on. Further, assume that the behavior to be explained is forward motion. That is, the question to be explained is: Why does the car keep going? The driver answers, "This thing does it" (indicating the accelerator pedal). The driver would now hope to hear a simple "oh," indicating that the child was satisfied. Demonstration of the assertion would consist in showing that releasing the accelerator results in slowing the vehicle, while depressing it keeps it moving ahead. As stated, the driver has provided at best a partial explanation in the form of a sufficient model. It may satisfy the child--but it may not, and for good reason. Given the driver's "explanation," does the child really understand forward motion, even in the assumed context? A reasonable answer to this question is that the child does not understand it completely, for the simple reason that there exists more to be said on the subject. What has been expressed and demonstrated is one way of producing forward motion. What remains unexamined at this point is whether there may be other ways of producing forward motion, and if so, what all these methods have in common, aside from their common effect. The skeptical child might persist: What are the other things for? A demonstration that none of the other controls separately, or in combination, will produce forward motion would justify the conclusive and finally explanatory model: "Only this thing will make it go." In the given context, this is the last word on the subject. It gives not only sufficient, but also the necessary conditions for the behavior that is under consideration.

It is, of course, possible that the child may have assumed the assertion of necessary and sufficient conditions because of the driver's strong and singular emphasis on the accelerator. Misunderstanding is possible.

It is also possible that the child may indeed have been satisfied with the sufficient (but not necessary) conditions, in the sense of saying to herself: "Oh, if I put my foot on that thing, the car would go." The harried parent, assailed by a barrage of "why's," may well hope that models of sufficiency will serve. Our point here is that what is satisfactory for indicating how control may be exercised may be inadequate for the kind of logical explanation provided by full answers. Full explanation, then, is a sort of logical exhaustion. We take this as the ideal against which alleged explanations should be judged.

Let us take the example of the brick wall and examine necessary conditions in that context. We assume now that our wall builder sets out to answer: "Why is a brick supported?" He will quickly find that a brick does not need two bricks below it for it to be supported. (Remember, this is a very naive fellow.) In fact, if he persisted in the three-brick model, he might seriously delay his discovery of the necessary conditions for a brick to be supported by other bricks. So we take it that he begins experimenting in a reflective manner with just two bricks. With enough experience he might be tempted to conclude that whether brick A supports brick B has something to do with whether the middle of B hangs over an edge of A. He might then set out as a necessary condition for *any* brick (making an inductive leap) being supported is that its middle be supported, either directly by another brick, or indirectly by portions of itself which are in turn supported by other bricks. Thus he has arrived at a modest theory. That it does not speak of centers of mass is unimportant. It is quite satisfactory in his presently limited circumstances. Note the relative difficulty and uncertainty connected with developing necessary conditions.

1.6 Mulligan Stew Modeling

A modern trend seems to be a push for sufficient models. Computers make it relatively easy to build and exercise complicated models. Since the archetypical procedure in getting sufficient models consists in making sure that nothing remotely connected with the circumstances to be modeled is left out of the model, this approach might be called Mulligan stew modeling, in honor of the traditional hobo concoction. As has been remarked, in many circumstances, such an approach may not be ill advised. Producing a good model can be a very significant achievement. An accurate model of the national economy would be extremely useful, even if it provided no insight at all into the internal working of the economy. (Indeed, such a model could be used in building up a theory of the internal machinery. That is one of the possible productive uses of the model.)

On the other hand, aiming blindly for a sufficient model first in a basically theoretical study can be deceptively attractive. Mulligan stew modeling may appear attractive because of the superficial plausibility of the idea that if a sufficient model can be devised, the necessary conditions are necessarily included in the sufficient model. Unfortunately, this is not strictly true. Even if the necessary conditions are in a logical sense included in the stated sufficient conditions, they may be included in much the same sense in which it can be said that a beautiful statue is included in every block of granite. In our brick wall example, the necessary conditions were only implicitly given by the sufficient. If one knows nothing about gravity and how its effects on bricks can be predicted, it will not necessarily be obvious that if two bricks under the ends of another brick will hold it up, then one brick under the middle of it will do so as well.

On this last point, take a further example. Suppose that our problem is to explain what an automobile is, using the context of automobiles in use on present day roads, and how they can be identified. The problem then is to provide necessary and sufficient conditions for "automobileness." A simple sufficient condition is: "If it is a Buick, then it is an automobile." The important question is: Where in this sufficient model is a necessary condition to be found? The answer here appears to be, nowhere. That is, the necessary conditions for being an automobile do not exist in this model at all. Thus, more generally, necessary characteristics necessarily exist only in the thing or things modeled. They may exist in or among a sufficient model's properties, but unfortunately, they may not as well. That is, sufficient models do not necessarily "include" necessary conditions at all, certainly not if by "include" we mean "available through a search confined to the abstracted content of the model."

What the simple Buick condition suggests, of course, is that in continuing our search for necessary conditions, we might well want to examine real Buicks, since, if our model is correct, each Buick-object is an automobile. On inspection, these objects share some features, but differ in others. By definition, what is necessary in being an automobile lies in the shared features. But note that a search among these particular objects would not allow a decisive conclusion. The common features will, of course, consist of the properties that are necessary to be both an automobile and a Buick.

To complete the automobile example, what is required is to finish the search for commonalities. Here, that means that the search must be extended beyond the properties of the objects modeled by the one sufficient condition discussed. We want to look at more than Buicks. We want, ideally, to look at all automobiles in the modeling context. In theory, one way to do this is to consider all possible sufficient models. All possible sufficient models would include the set of properties that points to each individual automobile (e.g., the sets of serial numbers that characterize existing individual automobiles). This, of course, would usually be an unworkable tactic. More useful would be smaller numbers of sufficient conditions that specify groups of automobiles. That is, the search for necessary properties would be conducted using a set of sufficient models that workably specifies the complete range of automobiles appropriate to the present modeling context.*

To sum up this discussion, the intuitively compelling argument that *any* sufficient model provides necessary conditions is just not true. To think that it does is to confuse models with things modeled. To be sure, a sufficient model may provide necessary conditions, but these may well have to be searched for laboriously, and the search in turn will involve reference to the things modeled. Further, examination of a single model naturally restricts the sway of the conclusions that can be drawn regarding necessary conditions. Finally, the search for necessary conditions is importantly associated with discovering commonalities.

Mulligan stew modeling, then, has something to be said for it in connection with theorizing. In throwing everything into the modeling pot, one desirable outcome is the tendency to include a large number of sufficient

---

*Levins has emphasized the importance of dealing with sets of models in theoretical work. His ideas will be examined in more detail below.

conditions. This tends to generate larger numbers of relevant "objects" and enlarges the relevance of any successful searches for necessary conditions. Unfortunately, necessary conditions are not the end of the story. A complete explanation accounts also for what might be called the sufficiency structure. In the driving example, the accelerator was the single sufficient condition for forward motion. It was for that reason a necessary condition as well. In the brick wall example, a complete theory would examine in what other particular ways a satisfactory wall might be built. In general, specifying the sufficiency structure will require (1) an empirical search seeking sufficiency conditions which differ from those already found, and (2) a search through the factors explicitly given in the model for possible empirical (or logical) inclusions of these factors.

The second search mentioned above seeks to answer such questions as: Are all these conditions jointly required for sufficiency, or are they perhaps individually sufficient? This type of search deals with all single factors, all pairs, all triples, and so forth. It clearly becomes a formidable undertaking if the number of factors is at all large. (To conduct the search completely, all *levels* of all factors must be considered, if factors exist in degree.) For example, assume that model L consists of factors A, B, and C. Let L be a sufficient model. Suppose that factor C is removed; call this smaller model S, and S is found to be sufficient. Then--provided that factors A and B are not contained within C in some hidden way--factor C is assured to be unnecessary. But it remains open whether A and B must both be present for sufficiency, or whether one factor alone might suffice: A alone might be individually sufficient with B necessary or unnecessary, or vice versa. (Indeed, factor C itself might be individually sufficient. This possibility must be guarded against as well.)*

The sufficiency structure is important in explanation. As an example, assume that the sufficiency structure for model S is: A and B are jointly sufficient. Now if *no other sufficiency conditions exist*, they are necessary conditions as well. If a sufficient model's fit is good, it is tempting to assume joint sufficiency and exclusive representation by the model's factors. This does away with the search for the sufficiency structure, and provides a theory, since all factors in the model are necessary. As we have seen, this conclusion is only as good as the assumptions that support it. And in general, these assumptions ought to be supported by firm knowledge of the sufficiency structure. Unfortunately, in a large model, the number of conditions raises the cost of the search for the sufficiency structure extravagantly--so much so that often enough the search is not even begun.

1.7 Stone Soup Theorizing

Is there some other way of constructing an explanation other than by starting with a sufficient model, and then facing the possibly formidable task of discovering the sufficiency structure and then teasing out the necessary conditions? Could we not from the outset build models that emphasize necessary characteristics? The answer is

---

*The search for the sufficiency structure is at least approximated in practice by what is called sensitivity analysis. In sensitivity analysis variation in levels of individual factors is examined for its effects on fit. Limited to single factors, it is for that reason incomplete.

that we can try, and the attempt may well be worth the effort. We can only attempt, though, for the obvious reason that we can be sure that the characteristics we use are necessary only by an after-the-fact examination for characteristics common to all "manifestations" of the phenomenon. In practice this approach to theory building starts with a conjectured explanation. The theorist uses any source material available, such as relevant theory or other logical or heuristic analyses which suggest what ought to be necessary characteristics of the phenomenon. In its extreme form, this approach builds "explanations" for phenomena that may not exist. This is "stone soup" theorizing. It tends to build very spare models, and often, naturally, to forfeit resemblance to the real world.

This is to recognize that stone soup theorizing faces difficulties just as does an approach via Mulligan stew modeling. Whereas "sufficient modeling" risks producing means of dealing with circumstances that do not contribute to our understanding of the circumstances, "necessary theorizing" risks producing knowledge of events for which the explanation is self-evident. And all this when the modeling process is successful in fitting the observables. When the process goes awry, Mulligan stew models are hodgepodges so expensive to test that their fit is never assessed, and stone soup theories end as empty exercises in logic.

Since our own recipe is closer to stone soup than it is to Mulligan stew, we devote considerable effort to interpretation in what follows. If our interpretations do not match reality satisfactorily, our hope is that they will nevertheless help explain aspects of phenomena that *do* match the model, and thus help to suggest better theories.

Another difficulty with this approach is equally

obvious. A model consisting of only the factors A, B, and C simply might not be a sufficient model even were A, B, and C *known* to be necessary. Sufficiency cannot be guaranteed. In particular, a sufficient model is not guaranteed by any certain number of necessary conditions being present. It might be argued that if one includes *all* necessary conditions, the model will necessarily be sufficient. Even neglecting the obvious point that if one is building a theory, one does not know how large the set of necessary conditions is, the assertion is incorrect. Referring to the automobile example, if one assembled a device that had a self-contained means of propulsion and other necessary conditions, one might have something that resembles each automobile but which would be recognized to be none of them. This construction will have all that our class of automobiles has in common, but will have none of the characteristics that our automobiles possess in addition to these common characteristics. For example, it will have no particular brand name, no particular paint color, and so on, and so will not be a particular automobile.*

The lesson is that a search for sufficiency is involved, just as in the converse case when necessary

---

*It will not be a particular automobile unless we allow our necessary conditions themselves to become arbitrarily complex, that is, of the sort that would say, "Each automobile shall have a particular brand name: (choose one of the following) Autocar, Buick, Cadillac, DeSoto. . . ." Clearly, this tactic distorts the notion of what a factor is to an unhelpful degree, and would not be acceptable in most cases of empirical modeling, since factors or conditions are expected to summarize data rather than to enumerate them. The extreme case of this tactic would be to set up as the one necessary condition: "An automobile has characteristics C(i), where C is the set of characteristics sufficient to identify automobile i." This can easily be seen to be vacuous.

conditions are being sought. In a genuine search, only heuristic guidance is available. For example, if the behavior to be explained can be pulled into components for which there are known or suspected necessary conditions, then an attractive first attempt at a presumably necessary (but only hopefully sufficient) model would, of course, include all those conditions. As another example, a modeling program that has produced an increasingly close fit as conditions were added would be very attractive.

1.8 Some Current Views on Theorizing

Despite the risks of stone soup theorizing, this style of modeling appears to have a clear advantage, at least in explanatory work, since it is biased toward simpler models. As our analysis has suggested, and Popper pointed out years ago, simple models are important in scientific work, and hence in theoretical study (Popper, 1959). What we described above as a modern tendency to build sufficient models has not gone uncriticized. Popper's view of science can be contrasted with earlier conceptions of it. Pre-Humean science was (we will only sketch these points of view) a relatively gentle gathering of facts, the building of a mosaic in which an occasional brilliant mind would discern a hidden pattern. Popper argues that science is much more, and properly, like a daily *High Noon* shoot-out between contending conjectures of reality. Important from our point of view is the fact that Popper stresses the place of systematic conceptions of reality in the form of explicit conjectures that should guide the actual process of scientific theorizing. From our point of view he is emphasizing explanatory models.

Among those who argue in the Popperian tradition is Platt when he asks that scientists use what he finds in the most vigorous branches of science: "strong inference" (Platt, 1964). Strong inference is, according to Platt, the deliberate use of inference in a way that maximizes the information gained from experiment. It requires the self-conscious and self-critical setting up of hypotheses with a thorough exploration, before the fact, of what will be concluded from any experiment planned. It requires that experiments be set up so that any outcome is productive.

Levins (1966, p. 423), writing for population biologists, points out that

> . . . even the most flexible models have artificial assumptions. There is always room for doubt as to whether a result depends on the essentials of a model or on the details of the simplifying assumptions. . . . Therefore we attempt to treat the same problem with several alternative models each with different simplifications but with a common biological assumption. Then, if these models, despite their different assumptions, lead to similar results we have what we can call a robust theorem which is relatively free of the details of the model. Hence not truth is the intersection of independent lies.

To Levins a robust theorem is an attempt to achieve statements that are true independently of their expression. It is at the same time an attempt to reduce a set of models to what is common and thus presumably central and indispensable, that is, it is an attempt to at least demark the factors that are necessary to produce some behavioral property.

It is worth noting again that the process of making sense of the intersection of several models is not a trivial detail in theorizing.

1.9 Nominal and Effective Structure

Given that we have a collection of things that affect one another, we shall call the connections between the things the systems structure. Control over the functioning of a system's parts can change what might be called the effective structure of the system. Ashby discusses this point at length (see Ashby, 1960, pp. 169, 174ff.), but it is not hard to convince oneself that it is so. For example, in a river system, a dam that is not discharging water effectively disconnects from the whole basin all sources upstream of the dam. It does so, of course, only while it is confining water. During that time, however, system connectivity is clearly altered in the sense of where water is actually flowing, as contrasted with the pattern of flow when the dam is allowing water through.

To continue the river analogy, in this book the term "structure" refers to the river beds themselves, not to the flows. More generally, in the present work "structure" refers to what might be called a system's nominal structure. In a switching net, nominal structure is the pattern given by the wires that connect the switches. It is clear that a system's nominal structure gives an upper bound on the system's effective structure. Nominal structure is a static description. In a dynamic system it is the effective structure that counts. To be more precise, it is the effective structure that counts at any given instant of time. The effective structure depends on three things. It depends on the nominal structure, of course, but it additionally depends on the functional regimes existing within the individual elements of the systems, and on the specific behavior the system is presently showing. That is, whether the dam is actually severing the connections with some downstream point depends on the river beds, of course, but also on whether the dam has a gate that is now closed, and on whether the river just below the dam is no longer carrying water.

1.10 The Architecture of Complexity

The heading of this section is the title of Simon's essay on "fully complex" systems (Simon, 1962). Simon examines complex systems that are hierarchical in structure--effective structure, to use the distinction introduced above. Each of his systems' elements is composed of a subsystem. It is a "boxes within a box" model. Each box on being opened shows a set of boxes. Each box, then, is both a system and a subsystem. The modified Chinese box analogy is intended to suggest the qualitatively different influence characteristics that join together a set of subsystems to form a system, compared with the interactions among the entities that form the subsystems.

Simon asks what behavioral properties, especially, what stability properties, follow from this arrangement. His conclusion is essentially that stable subsystems provide such economy of construction for whole systems that in evolutionary settings this structural paradigm would be expected to occur commonly.

To contrast Simon's approach with that taken here, we ask what desirable behavioral properties are obtainable given functional control, and especially given function control in systems that have relatively arbitrary structure. (As mentioned, our "structure" is nominal structure.) We shall argue that functional control of certain types can indeed be used to provide desirable behavioral properties in a great variety of specific structural arrangements. We might then join Simon in proposing that these functional control techniques could be used in building up larger systems in some hierarchical manner. Similarly, it might be expected that nature would have noticed these advantages and therefore might be found using them. This specific point will be discussed at greater length when we take up Kauffman's series of articles on the genetic control net.

One possible difference between Simon's analysis and ours lies in his requirement of explicit hierarchical structure and our relative indifference to structural forms. This difference is more apparent than real. We focus on nominal structure as being a simpler and more practical way of specifying structural arrangements. Nominal structures are used constantly in ordinary settings. "You report to W, and are in charge of X and Y" would be one way in which a nominal interaction structure might be phrased. Our approach is simply to take what we get when it comes to effective structure. We do not insist on certain forms of effective structure, but then, neither do we resist them if they occur.

The difference between Simon's perspective and that guiding the present work really has more to do with the locus of the analysis and its purpose than it has to do with effective and nominal structures. Instead of Simon's broad sweep of pyramiding hierarchies, our attention is directed within single systems, and principally there, to systems whose nominal structure is, to an important extent, arbitrary. Further, the present analysis examines how these single systems may be controlled, that is, how they might be treated so as to increase the likelihood of desirable behavior. Simon, on the other hand, takes the behavior of his subsystems, whether they are stable or unstable, for granted. He does not ask how the behavior is achieved.

Following Simon, we might expect nominal structure to be hierarchical in a large number of complex systems, since this arrangement provides an advantage in the aggregation of larger systems. Further, if nominal structure is visible, as it might well be in some physical systems where the possibility of interaction is signaled by nearness or by some obvious physical linkage, the hierarchies would not only be neat, but would tend to be evident as well.

Again following Simon, we would also expect hierarchical effective structure to provide some advantage in aggregation. But as Wimsatt (1974) points out, once aggregated and still in the grip of an evolutionary process, a neat and effective hierarchy would be at a disadvantage compared with systems aggregated of subassemblies that could incorporate mutually adaptive changes which improve overall system survivability. Such coadaptive changes could well involve boundary crossing interactions that disrupt the hierarchy. The neat hierarchy could, in the right circumstances, provide an internal environment in which natural evolutionary experimentation could proceed, leading to interactional complexity.

Hence, in general, we could expect neat nominal hierarchies of systems in realms where subsystems do not change, or change slowly if they do change, say, in physics, chemistry, or geology. We could expect less distinct nominal hierarchies in realms where change is easier, perhaps, in the life and behavioral sciences. In the latter areas, we would expect less reliance for overall system function to be placed on the structural form than on functional means of achieving it. That is, we would expect to be required to understand such systems not only by reference to nominal structure, but additionally by reference to functional arrangements that provide useful overall behavior. Possibly, but not necessarily, we would also see (effective) structural hierarchies.

Simon is not unaware of these considerations. He is at pains to argue that hierarchies, nominal or effective, have distinct advantages for aggregation. He points this out in the context of an assumption of subassembly stability, reasonably enough, since some relative degree of stability would appear to be necessary for aggregation. On the other hand, there would be an additional advantage available to systems aggregated of subassemblies that have functional properties that speed up subassembly time, or make subassemblies more probable. Holland (1962) suggests autocatalysis as one such property. In what follows, we shall discuss others.

In summary, there are realms, especially those in which interaction possibilities for elementary system parts can change over time, where the natural course of events would lead to arrangements that resist hierarchical decomposition. In these areas we can expect explanation to be difficult.

1.11 Organized Complexity

One of the main purposes of this essay is to consider how modeling complicated things can contribute to our ability to understand and manage them. As suggested above some areas seem especially complex. In 1948, Weaver discussed the topic of complexity and indicated a particular area in which complexity seems especially intractable. This area he found populated by systems showing what he called "organized complexity" (Weaver, 1948). Weaver also distinguished "problems of simplicity" from "problems of disorganized complexity." In the former, there are few variables, and relatively simple relations are found to obtain among them. In the latter, one finds many-variable systems in which many details do not matter, and where statistical relations are satisfactory ways of describing the dynamics of the systems. These two problem areas are those, according to Weaver, where science has made impressive gains in analysis. He suggested that some of our most important problems, and least remarkable progress, has occurred where "organized complexity" prevails. Such situations are those where the analyst is dealing "simultaneously with a sizable number of factors which are interrelated into an organic whole" (Weaver, 1948, p. 539). That is, our analytic tools fail us most notably in accounting for the properties of systems which have more than a few parts and in which individual parts may importantly determine what the whole system does. (See also Platt, 1961.)

In an effort to learn how these difficult systems can be studied, we shall suggest that switching nets are fully complex systems, which at the same time can serve as approximating models for a variety of real-world systems. Moreover, they serve as simple enough models to suggest directions for work aimed at obtaining full explanations.

1.12 The Observer

A question or two may remain with the reader. If fully complex systems are truly as difficult to deal with as has been claimed, how is it that we intend to study them? If these nets are "simple," in what sense are they really complex? The answer is that individually, these nets are not intractable, except in the sense of being very difficult in general to analyze formally. They are indeed easy to simulate. But this misses two points. First, not much is known about the behavior of classes of nets. Questions about the behavior of *classes* of nets with structures and switch characteristics already given are not of much interest in circuit engineering, where most of the work on switching nets has been done. Second, if one asks what behavior can be expected in classes of nets, one has implicitly set up a modeling context that both reintroduces an element of complexity into the problem and does it in such a way that it can be interpreted sensibly.

But the reader's objection is valid in one sense. Fully complex systems of the sort we consider *are* complex, given our available mathematical machinery and the mass of detail that might be important in their analysis. To handle these systems, they must be simplified. Our approach is straightforward. In all cases we embed our fully complex systems in contexts that allow them to be handled by standard statistical methods. One point of this essay is to show that this is worth doing. As noted earlier, our approach resembles that of classical statistical mechanics. Because of the sharp distinction between static and dynamic features of our model, the present approach might be more accurately called statistical dynamics.

The contexts mentioned are provided by introducing an observer. It is the observer's assumed inability to know certain details in any given system that provides the stochastic aspects of the approach taken here. To anticipate discussion in the interpretive sections, this tactic makes sense in modeling important aspects of situations that confront real-world managers of large systems.

To put it differently, if we know the characteristics of a given net we may be able, depending on these characteristics, to deduce its behavior mathematically; we can, depending on the size of the net, observe its behavior directly through simulation. But from our point of view, single nets are uninteresting. We simply do not care that this given single net does this particular thing. *Groups* of nets *are* interesting, in that they carry important implications for modeling.

To summarize, the observer enters for two reasons: to simplify the fully complex system* and to model some complicated real circumstances.

1.13 The Observer and Ensembles

According to the preceding section, the observer provides the statistical dimension of our modeling approach. What do we mean by this? We are drawing on the observation that ignorance means that uncertainty and variability exist in some area. From this we argue that conceptually a population of relevant "objects" exists as well. For example, suppose that all we can remember of the name of a person to whom we were recently introduced to is that it began with the letter "C." It seems clear that in trying to remember the name we are dealing in some sense with a population consisting of all names beginning with that letter. We could try to reduce the size of that population by using other facts that we recall regarding the name: Was it that of a man or a woman? Was it a long or a short name? and so forth. The observer can be said to generate these populations or ensembles by knowing some aspects of a situation, and by not knowing others. The variety allowed by the unknown factors varying among the fixed known factors provides the range in the ensemble. In our approach, the ensembles used are populations of nets.

We pointed out above that the introduction of an observer both simplifies the situation (by allowing us

---

*Foo has suggested such an approach in the following way: "Relatively unexplored are [simplification] procedures which convert stochastic to deterministic systems and vice versa, although the introduction of the ubiquitous [external or internal] random 'noise' is one aspect of such procedures" (Foo, 1974, p. 10).

systematically to overlook certain details in our analysis) and helps model some facet of the situation. But how does this occur? What is it that a population of nets is supposed to resemble in the real world? The answer is that our ensembles do not have to resemble the *things modeled*. We have argued above that a model is best understood as a triadic relationship involving the model, the thing modeled, and some human purpose for, or context in which the modeling is undertaken. The proper correspondence required of ensembles used in our modeling lies in their fitting the *modeling context*. For us, ensembles model not things so much as an appropriate observer's perspective. An ensemble reflects, and can be considered to be generated by an observer's disinterest in, or inability to deal with all of the pertinent details in some basic model. It is in the area of organized complexity that ensemble modeling can be expected to be especially prominent.

Ensembles of nets promise some "fit" in several broad contexts: (1) where it is important to explain large scale behaviors which can be seen to be insensitive to structural or functional detail, (2) where it is important to understand structural properties common over groups of large systems, and 3) where control of the behavior of a large system must be exercised, or understanding developed in conditions that forbid the use of complete knowledge of static details.

1.14 Controlling Nets' Behavior via Small-Scale Characteristics

The control schemes whose efficacy we will examine consist in broad, system-wide influence *on small-scale or local features of the nets*. The manager (who is also, clearly, an observer of the system), in addition to not knowing certain details of the systems, is not allowed to alter or manage them in detail. The idea here is suggestive of one possible view of mass advertising: advertisers may not so much attempt through advertising to influence the identity of individuals' influence sources, but rather to change the way they act on the information they receive. Another analogy suggests how some drugs may affect behavior: by changing the functioning of neurons in a part of the nervous system rather than by changing the specifics of interconnections among the neurons.

We take it that real-world control includes control exercised in a system-wide "broadcast" manner. We ask what kind of theory follows if that control is aimed primarily at changing local details, not explicitly at affecting structure, or other system features, in the large.

Two kinds of control will be considered. The observer will be allowed to control the number of inputs the elements in the systems have, and will also be allowed control over certain general functional characteristics of the elements. It is only in our examination of the effect of input control that the observer will be assumed able to do anything to, or know anything about, a system's structure.

This attention to the effects on overall system behavior of small scale system characteristics is a central focus of the present discussion. The point is to understand what dynamic properties of a complicated whole can result from specifications that refer only to local and hence (presumably) more readily knowable details of the system.

1.15 Once Again, Why Simple Nets?

From a programmatic point of view, we will be following Weaver's charge to examine "fully complex" systems:

many-variable things in which individual variables can be important in behavior, and where the interactions among the variables is complicated.

We argue that rather simple switching nets can theoretically be attractive examples of fully complex systems. Their simplicity, while raising questions as to how well they resemble reality, should help provide relatively efficient guidance for theoretical development. Moreover guidance, at least of a negative sort, can be had even if the resemblance is not good.

For example, suppose that a very large real switching net is available to us for experimentation, and that our job is to build a functional model. Suppose also that it is very difficult to determine details of the real net, but that we are convinced that we have most of them. That is, suppose that we are convinced that our model net's a priori specifications describe the real net rather well. On experimenting with the real net, however, we find that the model's behavior does not fit well enough. This is not uncommon in theoretical work; the investigator likes the model, but its behavior does not match reality. What guidance in the development of theory is available here? Fairly obviously, we will tinker with our model net, aiming to modify its behavior suitably. That is, we will dig for more detail concerning the real net, and incorporate this into the model, making it more realistic in the static sense (and hoping to see these changes improve the dynamic fit).

In our modeling, the following is also possible. The model is known not to provide a complete fit in the static sense in some particular area of description, but its behavioral fit is reasonably good. In this case the suggestion is that *at least some of the unrealism may be irrelevant*. For example, suppose that for some specific real-world situation agreement can be had on a switching net model's specification, except one. Let that specification, for the sake of argument, be the binary nature of the net elements. Now suppose that the (binary) net model's behavior shows reasonable fit with salient dynamics of the real thing. Assuming that the number of states in elements is logically independent of the various other specifications, we could conclude that ternary--or n-ary--elements are not theoretical necessities for that modeling context. Depending on circumstances, such a conclusion could provide important new information about characteristics necessary to explain the real thing.

References

Ashby, W. R. (1960). *Design for a Brain*. (2nd ed). New York: Wiley.

Burks, A. W. (1975). Models of deterministic systems. *Mathematical Systems Theory* , *8*, 295-308.

Caswell, H. (1976). The validation problem. *Systems Analysis and Simulation in Ecology*, *4*, 313-325.

Coombs, C. H., Raiffa, H., and Thrall, R. M. (1954). Some views on mathematical models and measurement theory. *Psychological Review*, *61*, 132-144.

Foo, N. Y. (1974). Homomorphic simplification of systems. Technical Report 156, Computer and Communication Science Department, University of Michigan.

Holland, H. H. (1962). Outline for a logical theory of adaptive systems. *Journal of the Association for Computing Machinery*, *9*, 297-314.

Kauffman, S. A. (1974). The large scale structure and dynamics of gene control circuits: An ensemble approach. *Journal of Theoretical Biology*, *44*, 167-190.

Levins, R. (1966). The strategy of model building in population biology. *American Scientist*, *54*, 421-431.

Murphy, G. (1967). Pythagorean number theory and its implications for psychology. *American Psychologist*, *22*, 423-431.

Platt, J. R. (1961). Properties of large molecules that go beyond the properties of their chemical subgroups. *Journal of Theoretical Biology, 1*, 342-358.

Platt, J. R. (1964). Strong inference. *Science, 146*, 347-353.

Popper, K. R. (1959). *The Logic of Scientific Discovery*. New York: Basic Books.

Schnakenberg, J. (1977). *Thermodynamic Network Analysis of Biological Systems*. New York: Springer-Verlag.

Simon, H. A. (1962). The architecture of complexity. *Proceedings of the American Philosophical Society, 106*, 467-482.

Simon, H. A. (1973). The organization of complex systems. In H. Pattee (Ed.), *Hierarchy Theory*. New York: George Brazilier.

Simon, H. A. and Newell, A. (1956). Models: Their uses and limitations. In L. D. White (Ed.), *The State of the Social Sciences*. Chicago: University of Chicago Press, pp. 66-83.

Stevens, Peter S. (1974). *Patterns in Nature*. Boston: Atlantic-Little Brown.

Weaver, W. (1948). Science and complexity. *American Scientist, 36*, 536-544.

Wimsatt, W. C. (1974). Complexity and organization. In *Proceedings of the Philosophy of Science Association, 1972*, Dordrecht, The Netherlands: D. Reidel, pp. 67-82.

2

Formalizing the Modeling Framework

2.1 An Ensemble Approach

The initial chapter has provided a philosophical introduction to general systems modeling. Our notion of a model has been explicated; our notion of complexity has been clarified. It is the purpose of this brief chapter to formalize a methodological framework for such a modeling enterprise with, in particular, its probabilistic aspects and its inferential possibilities. It has been called (e.g., Kauffman, 1974) an ensemble approach. This approach is particularly well suited to the switching net models which are the focus of this discourse.

Conventional wisdom evaluates the effectiveness of a model according to "goodness-of-fit"-based criteria. As we shall see, the level of abstraction of our models and the lack of quantitative information about the original preclude such an approach in our context. Rather, the "best model" in the sense of resembling the original is an arbitrarily close duplication of the original. However, as we have argued in Chapter 1, if a model becomes virtually indistinguishable from the original, it acquires all the complexity of the original and becomes useless in explaining the behavior of the original. Hence we seek something less from a model--

that it specifically resemble only certain facets of the original. Of course, there can be many models which specifically resemble a particular facet or set of facets of the original. This immediately leads to the notion of an ensemble of models. In fact, the ensemble approach will take this collection and define (sub)-ensembles within it. More concretely, suppose that for the original or phenomenon* which we seek to model we have a reasonable conception of its local detail (this is the observer's contribution) but have insufficient time to or no possibility of detailing its overall complexity. We hope that in developing models to reflect typical local detail, we will learn more about the overall behavior of the original and thus that we will fall usefully between the extremes of "stone soup theorizing" and "Mulligan stew modeling." We recognize that in formalizing an ensemble of models for a phenomenon according to local detail, we are conceptualizing only a subset of all available models. We hope that at least some of these will explain fairly well the complexity of the phenomenon. As we shall discuss later in this chapter and again in Chapter 5, it is not unreasonable that models built from local detail approximating that of the phenomenon will exhibit overall behavior which resembles that of the phenomenon.

Why and how do we create "ensembles" from the collection of models we conceptualize as meeting local detail? (We henceforth assume that only models satisfying the prescribed local detail are considered.) For the former question, the full collection will be enormously large and may be expected to span too broad a

---

*In this chapter "original" and "phenomenon" will be used interchangeably. The latter suggests a more behavioral orientation, which is our interest.

set of overall behaviors, many of which are infeasible relative to the original. Through partitioning of this collection, we attempt to define ensembles such that, from ensemble to ensemble, behavior is "different." Hopefully, it will then be easier to identify ensembles whose overall behavior resembles that of the original. As to the latter question, for the models considered in the subsequent chapters, equivalence classes arise as a result of structural specification and/or functional control having natural interpretations as static descriptors of the local detail of the phenomenon. These equivalence classes define our "ensembles." If we have a collection of ensembles and if we know something of the overall behavior of the phenomenon, we have an implicit suggestion that some ensembles are more probable than others. But how may we discern this, particularly if (as will be the case in the applications we envision) we lack the adequate quantitative description of the overall behavior of the original needed actually to "fit" good ensembles? We can, instead, run through the collection of ensembles describing, for each, "typical" ensemble behavior. An equivalence class of models will itself usually be very large, so that exhaustive examination is impossible. Rather, we must sample from each to obtain "typical" behavior. Fortunately, the number of classes is usually reasonable in number--indeed, few enough to enable complete examination of the classes.

In examining typical ensemble behavior we may draw random samples and employ customary statistical description: for example, the empirical distribution, measures of centrality (mean, median, mode), and measures of dispersion (variance, range). We then conclude (without resorting to formal inference procedures) whether the equivalence classes thus defined result in differing typical behavior from class to class. If such differences do not arise, we may conclude that the classes as defined are ineffective and seek new class definitions. If they do, we might further hope that these differences can be ordered by the levels of structural specification and/or functional control. In any event, we may seek to identify ensembles whose typical behavior most resembles that of the original. We may, in fact, conclude that no ensemble offers behavior which agrees sufficiently with the original. In this sense our ensembles will be satisfactory only if they produce a rich enough diversity of behavior across the levels of structural specification and functional control. The issue here may be expressed succinctly as follows. We accept the fact that the unknown detail of the original accounts for within-ensemble variation. We need the static descriptors of the local detail of the phenomenon to create ensemble variety.

Continuing in this spirit, suppose that our ensembles are adequate. If we wish to modify the behavior of the original in a particular way, this suggests seeking ensembles whose typical behavior is modified in the appropriate direction. In fact, we might invert our perspective, boldly concluding that the scope of behavior available among our ensembles describes the scope of behavior available to the phenomenon within the presumed local detail. We might further assert that the sorts of behavioral changes arising through changes in levels of structural specification and/or functional control or through perturbation may be expected to occur to the phenomenon if the corresponding modifications are applied to it. In this sense we may posit means for achieving certain desirable behaviors in the original and also its response to certain types of external shocks or perturbations.

The preceding two paragraphs articulate the essence of the ensemble approach. The last sentence of the preceding paragraph expresses our greatest aspirations for such a methodology. If we are only moderately successful, if we only somewhat enhance our comprehension of complex behavior, we will still have developed a useful modeling framework.

Although the discussion above may seem vague and fuzzy (and the authors a bit fanciful), we assert that by utilizing switching net models (developed in the next chapter), we can formalize these ensembles. We shall see that these models exhibit "complex" behavior (intensity, stability, replication, etc.). We shall see that these models can be constructed solely by the specification of typical local detail in the form of levels of structural or functional controls, allowing ready partitioning into ensembles. We shall see that the ensembles thus defined vary considerably in their "typical" behavior. Most important, we can conceive of several interesting complex real world systems whose local detail "agrees" reasonably well with that assumed for these models. The applications include management in complex organizations, dynamic behavior of genetic control systems, and word-of-mouth dynamics in consumer markets.

We emphasize again the fact that our thrust in this modeling exercise toward finding good models is quite different from the customary "parametric" goodness-of-fit approach. There is no conception of formal "optimality" criteria, there is no effort to find a "best" choice. The modeler may hope to predict broadly the response of the phenomenon to varying local details, but there is no expectation of precise behavioral quantification. In providing simple specification of detail, the modeler accepts entire ensembles of plausible models. In providing sparse empirical evidence on the relationship between overall system behavior and local detail, the impossibility of selecting a best model or even a best ensemble of models is recognized. The inherent complexity of the system being modeled imposes this less mathematically rigorous estimation/prediction formulation on the modeler. Again the real objective must be kept in mind: to learn about the behavior of the ensembles in order to achieve a better understanding of the behavior of the original.

2.2 Relationship to the General Decision Theory Framework

To distinguish further an ensemble approach from more standard inference methodology, let us see what happens if we attempt to cast this approach in a general mathematical setting. Such a setting for statistical inference is the statistical decision theory framework (see Ferguson, 1967, Chap. 1). We describe this framework briefly and then demonstrate why our modeling approach fails to fall within it.

The statistical decision theory framework envisions a scenario as follows. It is desired to make inferences about some unknown characteristic or characteristics of the phenomenon. A space of models for the phenomenon is then formulated by allowing as possible models all conceivable or plausible values or levels of the characteristic or characteristics. That is, a model is equated to a point in a space of dimension corresponding to the number of characteristics of interest. The space is denoted by Θ and the models are indexed by θ. We would then assume that some model $\theta \in \Theta$, although unknown, is correct. We next take observations (data) on the phenomenon denoted by x belonging to some sample space X in an effort to draw inference as to which θ is correct. In order that the data allow inference about the true θ, it is assumed to have been drawn from a probability mechanism uniquely determined by θ. This is usually achieved by describing a family of probability distributions parametrized by θ and assuming that the data come from that (unique) distribution within the family corresponding to the true but unknown θ. The set of inferences available to us is usually formalized as a space of actions denoted by A. A particular action, $a \in A$, asserts something about the true θ: for example, that it is a particular point in Θ (estimation) or that it is in a particular subset of Θ (hypothesis testing). Typically, the data in conjunction with the parametric form of the family of distributions enable a plausible guess for θ. In selecting a "best" action, an appropriate optimality criterion (called a loss function) is introduced. The "best" action is chosen as that producing a suitable extremum for the criterion.

Now consider our ensemble-based approach. There is no difficulty in formulating the space Θ of possible models. In fact, Θ will be the set of ensembles meeting the local detail and θ will merely index these ensembles. However, we do not conceive of one particular θ as being correct. Several different values of θ could product comparable and plausible behaviors. Moreover, values of θ needed to specify an ensemble are only used to approximate the local detail of the original. It is unlikely that any particular θ means much in the original. As noted earlier, we do not anticipate microscopic examination of the original to assess this. If such an inquiry were undertaken, we would probably adopt a different modeling approach. Under the present approach, given θ, we select random (usually equally likely) models from that ensemble. Corresponding to each model sampled, we record the appropriate behavioral characteristic or characteristics, which then become the observation corresponding to that model. From the sample of observations we may compute the usual sample descriptors. The entire sampling procedure is merely simple random sampling from a finite population. By contrast, in the statistical decision theory framework, we sample from only one, unknown $\theta \varepsilon \Theta$ (that which is assumed to characterize the phenomenon!) and then attempt to infer which θ or set of θ's can most reasonably be assumed to have given us the sample we obtained. In the ensemble approach we sample at each $\theta \varepsilon \Theta$ and know precisely which θ is associated with each sample. In this sense, no θ is any more likely than another. Moreover, if we draw larger and larger samples from our ensembles, we learn more about each ensemble but nothing further about the original. As an action we would still intend to choose a good θ or set of θ's. But to infer one θ or a set of θ's as more plausible than the others requires information beyond the data drawn on the ensembles. Our basis for selection comes from knowledge about the phenomenon and this information is not a component of the framework. In contrast, the statistical decision theory framework assumes that the *only* data considered are those observed directly on the phenomenon. We have thus articulated the crux of the aforementioned "inversion perspective." We do not analyze a set of data to infer a good model; we analyze a set of models to infer a good (similar to that of the phenomenon) set of data (and hence a plausible model).

The reader might be tempted to suggest recasting the framework with x's as θ's and θ's as x's. However, this is immediately absurd. We do not conceive of one x as correct or best and we do not sample from

θ's at a given x. In fact, any attempt to recast our problem to fit the statistical decision theory framework is doomed to failure since our approach samples all ensembles but does not formally sample the phenomenon. The reader might thus suggest that what is really needed is to take the original, record some behavioral data, and then select the ensemble of models whose typical behavior is "closest" in some sense to that of the original. As stated earlier, we probably would not advocate an ensemble approach if this were feasible. But for the intended application, our knowledge of the overall behavior of the original and/or the potential for readily obtaining additional knowledge is limited. The extent of what we know in a global sense may be that it has an enormous number of components, that there exists some connectance structure between the components, that it carries out its activities in a reasonable time frame, and that it is rather stable against shocks or perturbations. Our best empirical insight may be at the local level. Thus organized complexity (as we have defined it in Chapter 1) precludes the precise quantification needed for formal model fitting.

Our approach is intentionally nonspecific in relating models to the phenomenon. We are interested in inferring behavioral patterns for the phenomenon from behavioral patterns of suitable ensembles. Certainly, such inference must be viewed as somewhat tenuous. Yet such extrapolation is a primary objective of our modeling approach. Why might we expect such extrapolation to be valid? The fact that the original operates and survives over time suggests a straightforwardness in its composition and construction. If we mimic this local simplicity in our models, it is plausible that we will achieve a degree of behavioral mimicking through our models.

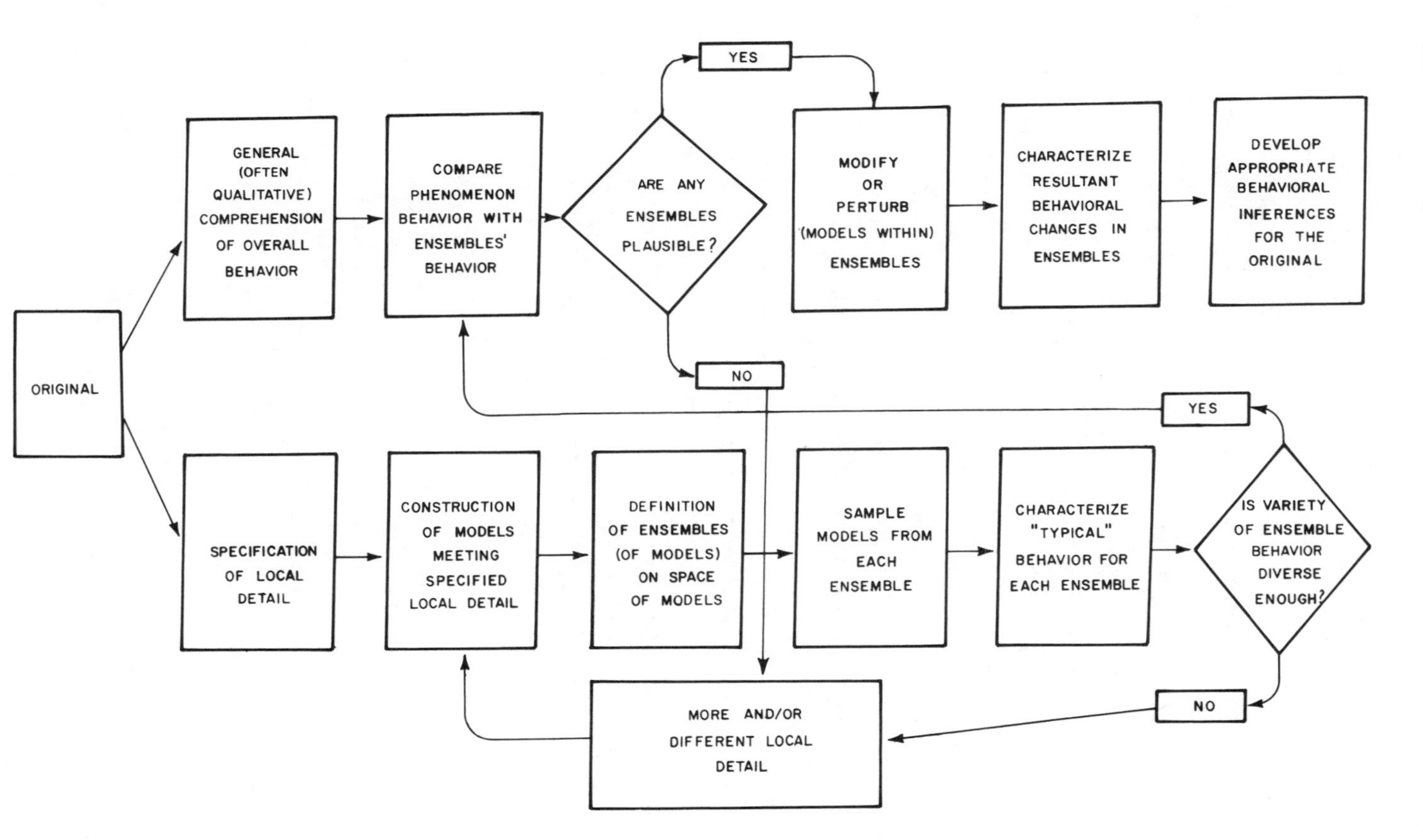

Fig. 2.1 Flowchart of an ensemble-based modeling approach.

In summarizing this chapter, a flowchart sequencing and detailing the components of the ensemble approach may be useful. Figure 2.1 provides such a chart. We also reiterate the basic parallels between a phenomenon (system) and models which, if reasonably well satisfied, render the models useful in interpreting and theorizing system behavior.

1. Whatever system behaviors are to be studied, for any individual behavioral characteristic, we presume that the variety of system behavior is "approximately spanned" by the total set of ensemble behaviors.
2. Specific local system detail can be "approximately interpreted" through levels of structural specification and/or functional control which define subensembles.
3. These subensembles offer diverse, distinguishable typical behavior such that at least some subensembles "agree" with typical system behavior.
4. Change in local system detail to effect particular change in typical system behavior "corresponds" to transition among ensembles to achieve comparable change in typical ensemble behavior.

References

Ferguson, T. (1967). *Mathematical Statistics: A Decision Theoretic Approach*, New York: Academic Press.

Kauffman, S. (1974). The large scale structure and dynamics in gene control circuits: An ensemble approach. *Journal of Theoretical Biology, 44*, 167-190.

3

Switching Net Models

3.1 Introduction

For the balance of this book we focus our attention on switching net models. Such models are simple in conception but offer considerable behavioral diversity and complexity. In this sense they become attractive models for examination. The former aspect allows them significant receptivity to mathematical analysis and to computer simulation, while the latter enables their applicability in the general systems sense to a variety of problems. We have three appealing applications of these models to offer. However, to appreciate the analogies fully requires some introductory development of these models. Such a development begins this chapter, with discussion of the applications following. In the subsequent two chapters we explore the behavioral details of these models from a theoretical and an empirical point of view, respectively. In the final chapter we interpret this analysis with respect to our application. Our discussions will regularly find us elucidating and drawing upon the modeling notions of the initial chapter. They will also reveal how one is naturally led to the employment of the ensemble approach of Chapter 2 in studying the expected behavior of these models. Hence the study of switching net models affords a microcosmic view of the various components involved in general systems modeling. Additionally, it is hoped that the unification attempted in this book with respect to switching net models will provide an impetus for further organizational activity in general systems research.

3.2 Basic Definitions

We begin by briefly introducing some terminology and notation. For definitional purposes we take a binary switching net to be a network with an associated set of Boolean transformations. In our work we assume no external inputs to the net. Strictly speaking, this defines an autonomous switching net, but henceforth we suppress the word "autonomous." Network models, in particular, randomly connected network models, have a considerable literature. Since one may envision the flow of mass, energy, or information through such networks, these models have been applied in such diverse areas as structural chemistry, sociology, and information retrieval. However, the widest area of application has been in biology, where examples include neural networks, the spread of excitation in cardiac muscles, the spread of contagious disease in a population, and the spread of cancer in an organism. A few illustrative references are provided at the end of this chapter (Alben and Boutron, 1975; Bell and Dean, 1972; Blumenson, 1970; Doreian, 1974; Rashevsky, 1960; and Stubbs, 1977). But it is not our purpose here to investigate this large body of published material. That task in itself can provide more than a monograph. We intend to focus on switching net models and as a result will draw from the literature only that which is pertinent to the study of these models.

Formally, a network is the couple (N,K) consisting of a set of N nodes or elements and K, an N x N connections matrix indicating the connections between elements. The entries in K are only 0's and 1's. A 1 in cell (i,j) indicates a connection *from* element i *to* element j, while a 0 indicates no connection. More conventional prose refers to a connection *from* an element as an *output* and a connection *to* an element as an *input*. A 1 at any diagonal element indicates "feedback" that is, the corresponding element draws input from its own output.

We now distinguish two terms often used interchangeably in the literature. We employ the word "connectance" to describe direct connection from one element to another and say that two such elements are "directly connected." In contrast, we take the word "connectivity" to describe eventual connection; that is, one element may be reached from another through a sequence of connections involving other elements. We would then say that the two elements are "connected." More precisely, we say that element i is connected to element j in p steps if the (i,j) entry in K^p (usual matrix multiplication) 1. Obviously, then, "connectivity" includes "connectance" and "connected" includes directly connected since the latter term in each instance corresponds to the case p = 1.

Returning to K, we define the sum of the elements in the i-th row to be the output connectance of the i-th element. Similarly, the sum of the elements in the i-th column of K is called the input connectance of the i-th element. The sum of all the entries in K divided by N is the average connectance (input or output) of the network. Constant input (output) connectance k means that each column (row) of K has the same k. A network is homogeneous in connectance if it has constant input and output connectance; that is, each element would have as many direct connections from itself as it has to itself. A network is fully connected if k = N. In the case of input connectance = output connectance = 2, one may formulate a "neighbor-connected" network which may be given a planar lattice representation (see Atlan et al., 1982). Output connectance is natural to consider in the study of neural nets since it indicates the number of axons sent out by a stimulated neuron. However, in switching nets, input connectance is more crucial since the definition of the Boolean transformation which determines an element's responses requires specification of the number of inputs (and, in fact, a labeling of these inputs). In the switching nets we study, we usually specify a constant input connectance k leaving the output connectance to vary from element to element. Figure 3.1 offers a graphic depicture of a network with N = 4 and K given as well. Note the constant input connectance, k = 2.

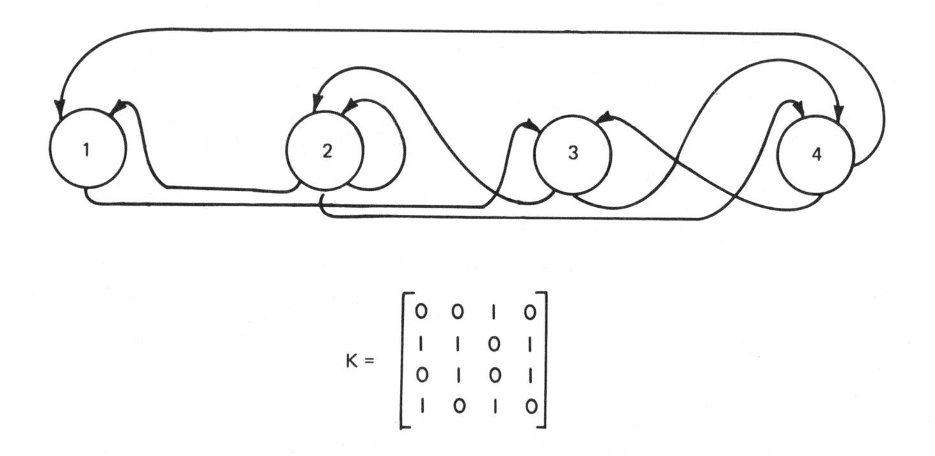

Fig. 3.1 Network with N = 4 and associated connections matrix X.

The study of networks entails either a deterministic selection of K which leads to the study of classical graph theory or a probabilistic selection of K which suggests an ensemble approach for analysis. We restrict ourselves to the latter combinatorial approach partly because this sort of selection is in the spirit of our conception of random sampling of models from an ensemble of models as discussed in Chapter 2. Again, this conception arises from the fact that our applications imply the existence of a complex web of interconnections among network elements but not the knowledge of specific detail.

Such randomly connected networks were first studied in detail in the late forties by Rapoport (1948) and by Shimbel (1948) for the case of unit output connectance and in the early fifties by Solomonoff and Rapoport (1951) for multiple-output connectance. Further early work by these authors [Shimbel (1951), Solomonoff (1952), Rapoport (1954)] elaborates this seminal development. A survey article by Stubbs and Good (1976) notes that since that early effort there have been numerous papers discussing various aspects of connectance and connectivity in random networks. The work covers material dealing with such areas as ecosystems, metabolic networks, neural nets, and general theoretical development.

Structure is for our purpose the critical concept in discussing switching net models. Connectance in this sense becomes the most important aspect of a network. Rapoport observed that on the simplest level there is an isomorphism between the graph of a network and the connections matrix of a network (Fig. 3.1 clearly illustrates this notion). We abbreviate this idea in asserting that the network itself is isomorphic to its connections matrix. The connectance structure of a network in our models will be assigned via a random selection mechanism constrained only to a constant input connectance k. Hence the choice of k represents the only structural control we can exercise over the network.

But a switching net model is more than a network. We have defined it to include an associated collection of Boolean transformations, one for each element of the network. It is in the selection of particular Boolean transformations from the collection of all Boolean transformations that we achieve diverse and intricate behavior for the models. These various types of constraint in the choice of Boolean transformations allow interpretable and sophisticated control to be exerted on the models.

We offer a brief perspective on the switching net literature. The study of switching nets was initiated by McCulloch and Pitts (1943). They defined a formal neuron and proposed its use to describe real neural interaction in the central nervous system. Subsequent researchers influenced by this work have concentrated on models of subsystems of the brain and other neural phenomena in which "all or none" response of the elements is appropriate. Arbib (1972) presents a review of much of this work. We do not devote further attention to this area of application since it does not formalize the Boolean transformation feature of these nets. The Boolean transformation is present but not discussed; rather, the focus is on the equivalent (as we shall see) state diagram. Thus the aspect of functional control available through restrictions on the transformation which we are able to interpret and analyze usefully is not emphasized (although threshold levels, an example of such control, are implicit). Kauffman (1969, 1970, 1974) first applied switching nets to the study of genetic systems. He highlights the sort of functional control we are concerned with and we present his discussion interspersed through this chapter and Chapter 4. He also presents a substantial behavioral analysis based primarily on simulation which has been extended by Sherlock (1979a,b). We defer discussion of this material to Chapter 5. The authors in a series of papers have described applications to management strategy and to advertising policy again from a functional point of view. Caianiello (1973) and Cull (1971) have studied these nets with the purpose of developing an attractive linearization of switching nets, thus enabling a matrix calculus for these nets amenable to linear algebra methods. Earlier work of this sort was begun by Latour (1963). This linearization technique is useful in calculating certain behavioral characteristics of these net and will be discussed in Sect. 3.4.

We now define a Boolean transformation. Such a function, which we often call a mapping, is a rule which, for an element with k inputs, prescribes an output value for each possible vector of input values. The inputs are binary valued as is the output. Thus the mapping m is denoted by

$$m: \quad X\{0,1\}^k \longrightarrow \{0,1\}$$

Clearly, there are 2^k possible input vectors and specification of m requires its value for each of these 2^k input vectors. Since the output is binary as well, there are 2^{2^k} possible Boolean transformations. Of course, one may readily conceptualize more than binary response for the elements in a network. If we allow n responses, we refer to our systems as n-ary (instead of binary) switching nets and this number becomes n^{n^k}. The inputs to an element will be ordered and labeled as $x_1, x_2, \ldots x_k$. It may be convenient at times to denote a mapping m as a function of its inputs, [i.e., $m(x_1, x_2, \ldots x_k)$]. If all N elements in a network are governed by the same mapping, we refer to the switching net as homogeneous, otherwise as heterogeneous.

A natural representation of a mapping is in tabular form with the input values arranged either in lexicographic of monotonic order. Table 3.1 illustrates the general representation of a mapping on three inputs with the input values in lexicographic order while Table 3.2 has the input values in monotonic order. Lexicographic order is more easily grasped from Table 3.1 than through formal definition. For k inputs the first 2^{k-1} rows will have $x_k = 0$ and the remaining 2^{k-1} rows will have $x_k = 1$. Also, the first 2^{k-1} rows will have $x_{k-1} = 0$, the next 2^{k-2} rows will have $x_{k-2} = 1$, the next 2^{k-2} rows will have $x_{k-1} = 0$, and the last 2^{k-2} rows will have $x_{k-1} = 1$. The pattern is clear and obviously all possible input vectors will be developed in this manner. Monotonic order arranges the input rows in terms of increasing number of 1's from top to bottom. This order is obviously not unique, but again it is apparent that all possible input vectors will be created.

Lexicographic order is more precisely defined and more prevalent in the literature, but selection of a particular order will usually be dictated by the particular form of functional control we are applying through the Boolean transformation. If we are dealing with forcibility, lexicographic order is most convenient; if we are dealing with threshold, monotonic order is most convenient; and if we are dealing with internal homogeneity, either is convenient. (These types of control are defined formally in Chapter 4.)

Returning to Fig. 3.1, where, we recall a constant connectance k = 2 was imposed, suppose that we label the

Table 3.1 General Representation of a Mapping on Three Inputs with Inputs in Lexicographic Order

x_1	x_2	x_3	m
0	0	0	m(0,0,0)
1	0	0	m(1,0,0)
0	1	0	m(0,1,0)
1	1	0	m(1,1,0)
0	0	1	m(0,0,1)
1	0	1	m(1,0,1)
0	1	1	m(0,1,1)
1	1	1	m(1,1,1)

Table 3.2 General Representation of a Mapping on Three Inputs with Inputs in Monotonic Order

x_1	x_2	x_3	m
0	0	0	m(0,0,0)
1	0	0	m(1,0,0)
0	1	0	m(0,1,0)
0	0	1	m(0,0,1)
1	1	0	m(1,1,0)
1	0	1	m(1,0,1)
0	1	1	m(0,1,1)
1	1	1	m(1,1,1)

left inputs as input 1 (i.e., x_1) and the right inputs as input 2 (i.e., x_2). Then in Table 3.3 we specify a choice of four Boolean transformations associated with the four elements; that is, m_1 is associated with element i, i = 1, 2, 3, 4. Figure 3.1 and Table 3.3 taken together provide an example of a heterogeneous binary switching net model.

Table 3.3 Associated Boolean Transformations for the Networks in Fig. 3.1

x_1	x_2	m_1	m_2	m_3	m_4
0	0	0	1	0	1
1	0	1	1	1	0
0	1	0	1	0	1
1	1	1	0	0	0

The collection of Boolean transformations on k = 2 inputs have received special attention in the literature (see, e.g., Atlan et al., 1982 or Babcock, 1976). This is due to their simplicity, to their implicit behavioral ramifications (as discussed in the following chapters), and to their interpretation as familiar logic functions. To this last point, each of these 16 transformations expresses a logical relationship between the two inputs; for example, m_1 below expresses the notion of "and" and m_2 below expresses the notion of "exclusive or."

x_1	x_2	m_1	m_2
0	0	0	0
1	0	0	1
0	1	0	1
1	1	1	0

If, for example, m_2 is written in a "truth table" format, the logical function notion becomes apparent.

m_2		x_2: 0 (F)	x_2: 1 (T)
x_1	0 (F)	0 (F)	1 (T)
x_1	1 (T)	1 (T)	0 (F)

With all the model components in place, we turn next to a discussion of the behavior of such net models.

3.3 Net Behavior (Introduction)

In describing the behavioral characteristics of a switching net, we require some further terminology. The state of the net is an ordered (left to right) N vector wherein the i-th coordinate is the current output value of the i-th net element. The net progresses through states in a discrete manner. The state of the net at time t + 1 is determined from its state at time t; that is, the output values of the elements at time t become input values in determining the new output values of the elements at time t + 1. In this manner the net moves from state to state over time in a determinant fashion. It is apparent that there are but a finite number of distinct net states, in fact, 2^N of them. It is thus clear that from some initial net state the net must eventually come to a state that it had previously passed through. Doing so, it must then repeat the sequence of intermediate states. Such a sequence of states is called a cycle. Note that every net must have at least one cycle. The sequence of states from the initial one until entering the cycle is called the run-in. The number of distinct net states in the cycle is called the cycle length; the number of net states (necessarily distinct) in the run-in is called the run-in length. The sum of a cycle length and an associated run-in length is called the disclosure length. Note that the cycle length may range from 1 to 2^N, while the run-in length may range from 0 to $2^N - 1$.

For the switching net of Fig. 3.1 and Table 3.3, the reader may verify that beginning, for example, in state 1110, the net moves through the states listed in Fig. 3.2.

1110 → 0010 → 0101 → 1100 → 0110 → 0000 → 0101

Fig. 3.2 A state sequence for the net in Figure 3.1 and Table 3.3.

Note that in this sequence we have both a run-in and a cycle. The run-in length is 2 and the cycle length is 4.

Recognizing that we have described but one state sequence suggests that we might think of the entire collection of state sequences. In referring to this collection, which reflects the cyclical behavior of the net, we employ the term "cycle space" or "behavior space." More generally, in every switching net each state belongs to either a run-in or a cycle. Using Markov chain model terminology, we may describe these states as transient or recurrent (cyclic), respectively. The cycle space of the net would partition the state space of the net into distinct cycles, with each cycle having its associated transient states. Figure 3.3 exhibits the cycle space for our example. The lines connecting states are directed in order to sequence the states graphically.

As the number of elements in the net grows large, the number of net states increases dramatically. For example, even with N = 20 the number of states is already $2^{20} \approx 10^6$ while for N = 100 the number becomes enormous--$2^{100} \approx 10^{30}$.

Hence the lengths of cycles can grow large but, more important, it will not be feasible to detail the cycle space explicitly in these cases. Nonetheless, net models with large N are precisely the ones that we would wish to use in approximating complex real world phenomena. Hence we will still want to know in an approximate sense how the number of distinct cycles and the lengths

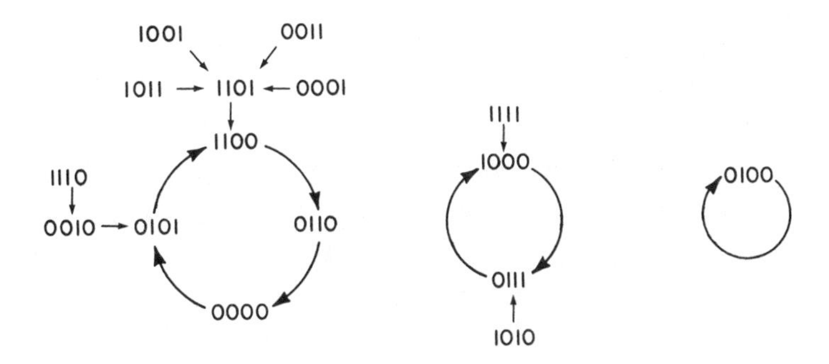

Fig. 3.3 Cycle space for the net in Figs. 3.1 and 3.2.

of these cycles depend on N. We will also want to know the effect on the cycle space of applying certain types of functional control through the Boolean transformations in the net. Difficult questions of this nature will be taken up in Chapter 4 and provide the motivation for the work contained therein.

In looking at the cycle space of a net, it becomes natural to examine the stability of the net's behavior. Suppose that as the net passes through states in a cycle, minor perturbation is introduced. We inquire as to whether

1. The net tends to return to the same cycle.
2. The net tends to return to a particular subset of the cycles in the cycle space.
3. The net is as likely to wind up on any cycle in its cycle space.

Case (1) represents the strongest form of stability. Case (2) is usually referred to as "restricted local reachability." From a modeling perspective, case (3) might render these nets less appealing in relating them to real-world phenomena.

Formally, we shall define a minor perturbation as changing the value of one element in a state on a cycle. After such a perturbation we would then note to which cycle the net returned. By perturbing each state on each cycle in all possible ways one element at a time, one may obtain a transition probability matrix called the flow matrix among net cycles. This is done by recording the total number of times the net returned to the perturbed cycle or ran to each other cycle. Let the rows denote the initial cycle and the columns the resultant cycle after perturbation. Then dividing the value in each cell by its row total yields the desired matrix of transition probabilities.

For our example, if we label the cycles in Fig. 3.3 from left to right as 1, 2, and 3, then Table 3.4 presents the corresponding flow matrix. This illustrative net would appear to have little stability and too few cycles to offer meaningful discussion of restricted local reachability.

As N grows large, theoretical examination of the stability of switching nets becomes difficult, particularly in the presence of the functional control. Empirical consideration of this problem is taken up in Chapter 5.

Table 3.4 Flow Matrix for the Net in Figs. 3.1 and 3.2

	1	2	3
1	0.5	0.25	0.25
2	0.75	0.25	0
3	1	0	0

Cycles are not the only behavioral characteristic of a switching net which one may usefully study. The choice of system variables to be monitored usually derives from the intended application of the net model. Often these will be cycle-related variables such as we have discussed, but other possibilities exist. For example, one may monitor the proportion (density) of 1's in successive nets states in an effort to detect whether the proportion tends to be maintained or whether a directional trend is exhibited. From these proportions one could compute the average behavior of the sequence and its variability. Such a collection of variables will be appropriate in our application to modeling advertising policy in consumer markets.

One may also compute distance variables between successive state vectors according to some appropriate distance measure. In this case the sequence of variables helps to discern whether states are becoming more similar or less similar over time. The sequence may also be utilized to attempt to detect whether a shock characterized by a change of Boolean transformation at the net elements has been applied to the system.

Of course, the fact that in an unperturbed situation any net sequence of states must eventually fall into a cycle implies that both the sequence of proportions and the sequence of distances will eventually be cyclical as well. In fact, because these variables reduce state vectors to one-dimensional observations, the latter sequences may exhibit shorter cycles. Does the fact that these variables are also cycle related deny their usefulness in modeling situations? The answer is no. With large net size, cycle lengths will tend to be quite large. Since we intend to apply frequent structural and/or functional changes to the model, in applications where these variables are appropriate it is unlikely that we will observe these sequences long enough to encounter cyclic behavior. The more crucial issue is to establish intervening control that, for example, stabilizes the sequence of proportions or decreases the sequence of distances.

Increasing the response possibilities for the net elements, say to n values instead of 2, leads to the examination of what we have called n-ary switching nets. Such nets offer a wider range of variables to study, but at the expense of much less tractability both behaviorally and analytically.

3.4 Equivalent Representations of Switching Nets

The cycle space or state diagram in Fig. 3.3, which was derived from Fig. 3.1 and Table 3.3, is, in fact, equivalent to Fig. 3.1 and Table 3.3. That is, from Fig. 3.3 we may retrieve the network diagram and the associated Boolean transformations. To see how this may be done, we refer to Table 3.5. In Table 3.3 at the left all 16 possible net states are arranged by row in lexicographic order. (It will be shortly apparent that any order may be used.) At the right we have four columns, corresponding to each of the four associated mappings. For a particular row (state), to obtain the entries for m_1, m_2, m_3, and m_4, we merely refer to the state diagram and find the next state and insert this state as m_1 through m_4. For example, at row 1101 we would find the next state to be 1100 and thus have $m_1(1,1,0,1) = 1$, $m_2(1,1,0,1) = 1$, $m_3(1,1,0,1) = 0$, and $m_4(1,1,0,1) = 0$.

At this point Table 3.5 describes a fully connected net. We need to clarify the connections matrix, which we may do as follows. To establish whether, say, element 2 is an input to element 3, we must ask whether $m_3(x_1,0,x_3,x_4) = m_3(x_1,1,x_3,x_4)$, for all x_1, x_3, and x_4 combinations. If the answer is yes, then the value of

Table 3.5 Reconstruction of Boolean Transformations from a Net Diagram

Elements				Mapping			
1	2	3	4	m_1	m_2	m_3	m_4
0	0	0	0	0	1	0	1
1	0	0	0	0	1	1	1
0	1	0	0	0	1	0	0
1	1	0	0	0	1	1	0
0	0	1	0	0	1	0	1
1	0	1	0	0	1	1	1
0	1	1	0	0	0	0	0
1	1	1	0	0	0	1	0
0	0	0	1	1	1	0	1
1	0	0	1	1	1	0	1
0	1	0	1	1	1	0	0
1	1	0	1	1	1	0	0
0	0	1	1	1	1	0	1
1	0	1	1	1	1	0	1
0	1	1	1	1	0	0	0
1	1	1	1	1	0	0	0

element 2 does not affect the resultant value of element 3; that is, element 2 does not provide input to element 3. If the answer is no, the converse is true and element 2 does input to element 3. In this way we could establish that elements 1 and 4 input to element 3, while elements 2 and 3 do not and thus reduce the mapping m_3 in Table 3.5 to just two inputs (4 rows), as in Table 3.3.

Without formalizing the foregoing process, it should now be clear that any binary switching net model is equivalent to its state diagram. For our purposes, in describing and analyzing these nets the former representation is more convenient. This is so by virtue of the previously expressed fact that the state diagram representation of the switching net subverts the Boolean transformation notion. However, the state diagram does lead us to other net representations and, in particular, to the clever linearization of Caianiello (1973) and Cull (1971), which we now discuss.

The state diagram may readily be converted to a transition matrix as follows. Identify with each state a standard basis vector in 2^N-dimensional Euclidean space; that is, state i will be isomorphic to the vector e_i which has a 1 in the i-th row and 0's in the remaining $2^N - 1$ rows.

Then clearly there is a $2^N \times 2^N$ matrix T such that $T_{ij} = 1$ if and only if state i is the successor to state j. T will be a matrix of 0's and 1's with exactly one 1 in each column. Clearly, there are $(2^N)^{2^N}$ distinct T matrices, corresponding to the number of mappings from a set of 2^N elements into itself, that is, corresponding to the number of state diagrams for a switching net composed of N elements.

Thus the matrix T is equivalent to a switching net model. As a simple example with N = 3, consider the state diagram

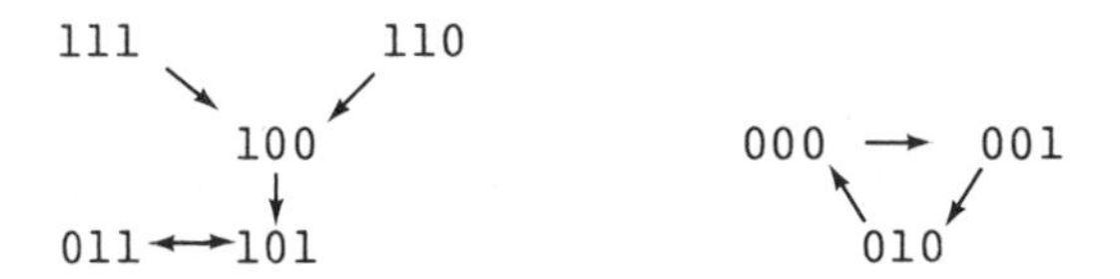

Then with the states arranged in lexicographic order, T becomes

	000	100	010	110	001	101	011	111
000	0	0	1	0	0	0	0	0
100	0	0	0	1	0	0	0	1
010	0	0	0	0	1	0	0	0
110	0	0	0	0	0	0	0	0
001	1	0	0	0	0	0	0	0
101	0	1	0	0	0	0	1	0
011	0	0	0	0	0	1	0	0
111	0	0	0	0	0	0	0	0

Next, we note that if $x^{(t+1)}$ and $x^{(t)}$ are N-dimensional vectors representing the state of the net at times t + 1 and t, respectively, then the switching net is equivalent to

$$X^{(t+1)} = F(X^{(t)}) \tag{3.1}$$

where F is the "next state" function. In this notation we have F denoted by

$$F: \; X\{0,1\}^N \longrightarrow X\{0,1\}^N$$

In general, the function F will be nonlinear. But it is also a mapping from a finite set to the same finite set. This suggests that F in (3.1) will behave like a permutation function, and since a permutation function has a matrix representation, there should be a matrix representation of F.

Before proceeding we note that the transition matrix directly provides a linear next state function. With the states arranged in lexicographic order, suppose that the i-th ordered state is represented as a unit basis vector in 2^N dimensional space, Y, having a 1 in row i. (In fact, any one-to-one assignment of states to basis vectors could be used.) Then

$$Y^{(t+1)} = TY^{(t)} \tag{3.2}$$

where $Y^{(\cdot)}$ is the state $X^{(\cdot)}$ in this representation. Of course, this linearization is not at all dependent on the binary character of the net models; it will work with any finite state system whose states are similarly represented. What we seek is a matrix which also linearizes but which is applied to states retaining their binary character and which reflects (in fact, through its rows, as we shall see) the Boolean transformations computed by each of the elements of the net.

To create this "function" matrix which we also denote by F, let us first recall some basic properties of the simplest finite field, the integers modulo 2. In this field there are only two numbers, 0 and 1. The two operations, addition represented by + and multiplication represented by juxtaposition, are defined by the relationships

$$0 + 0 = 1 + 1 = 0$$
$$0 + 1 = 1 + 0 = 1$$
$$00 = 01 = 10 = 0$$
$$11 = 1$$

A moment's reflection reveals that there are exactly 2^{2^N} functions of N variables over this field; that is, each of the 2^N points in the domain can be assigned one of two values in the range. This is merely a restatement of the fact that there are 2^{2^N} possible Boolean transformations with N inputs; that is, each function is equivalent to a Boolean transformation. More important, each function is equivalent to a multinomial in N variables over this field. To clarify the notion of a multinomial, let $x_1, x_2, \ldots, x_N$ denote the N variables. There are 2^N possible subsets of the x's which may be drawn. For the i-th subset, $i = 1, 2, \ldots, 2^N$, let g_i be the function which is the product of all the x's in this subset. (For the empty set, take $g = 1$.) Let c_i, $i = 1, 2, \ldots, 2^N$, be a set of constants each of which is either 0 or 1. Then $\Sigma c_i g_i$ defines a multinomial in N variables over this field. For example, if N = 3, we have

$$g_1 = 1, \quad g_2 = x_1, \quad g_3 = x_2, \quad g_4 = x_1x_2$$
$$g_5 = x_3, \quad g_6 = x_1x_3, \quad g_7 = x_2x_3, \quad g_8 = x_1x_2x_3$$

and if $c_3 = c_6 = c_8 = 1$ with the remaining $c_i = 0$, we obtain the multinomial

$$x_2 + x_1x_3 + x_1x_2x_3$$

Since there are 2^{2^N} distinct choices for the set of constants, there are 2^{2^N} distinct multinomials over this field. Any function f of N variables over this field may be represented as a 2^N-dimensional vector of 0's and 1's, with the 0's and 1's being the coefficients of the multinomial representing this function. The function or multinomial is then represented by an inner (dot) product on this field, that is,

$$f = \underset{\sim}{c}^T \underset{\sim}{g} \tag{3.3}$$

where $\underset{\sim}{c}$ is the vector of c_i's and $\underset{\sim}{g}$ is the vector of g_i's. The function can be evaluated at a particular point $(x_1,...,x_n)$ by inserting these values into g.

It is quite apparent that given a specific multinomial, one can readily find the equivalent Boolean transformation by inserting state vectors (points) into (3.3) in lexicographic order and obtaining the function values. However, given a Boolean transformation (i.e., the function value at each state), it is not at all clear how to obtain directly the equivalent multinomial. That is, how may we find the corresponding $\underset{\sim}{c}$ in a systematic manner?

Suppose that we order the states and the g_i's in the manner suggested by our examples with N = 3. That is, we place them in lexicographic order. This ordering is clear for the states. For the g functions, since there are 2^N of them as well, they may be placed in correspondence with the states where ordered function g_i takes as its subset those x's which are set at 1 in ordered state i. Now suppose that we derive a $2^N \times 2^N$ matrix G whose j-th column is g evaluated at the j-th ordered state, that is, a representation of the j-th ordered state through the multinomials. Note that an element g of $\underset{\sim}{g}$ evaluated for a given state s can only be 1 if every x in g's subset is valued at 1 in s. Hence by the defined order on the g's and s's, G_{ij}, which is the i-th ordered product g_i, evaluated at the ordered state s_j must be 1 if i = j and 0 if i > j. Therefore, G is an upper triangular matrix of 0's and 1's with all diagonal elements equal to 1. It may also be seen that G is symmetric *not* with respect to the main diagonal but with respect to the diagonal orthogonal to it. For our example, with N = 3, G becomes

$$G = \begin{array}{c} \\ 1 \\ x_1 \\ x_2 \\ x_1x_2 \\ x_3 \\ x_1x_3 \\ x_2x_3 \\ x_1x_2x_3 \end{array} \begin{array}{c} \begin{array}{cccccccc} 000 & 100 & 010 & 110 & 001 & 101 & 011 & 111 \end{array} \\ \begin{bmatrix} 1 & 1 & 1 & 1 & 1 & 1 & 1 & 1 \\ 0 & 1 & 0 & 1 & 0 & 1 & 0 & 1 \\ 0 & 0 & 1 & 1 & 0 & 0 & 1 & 1 \\ 0 & 0 & 0 & 1 & 0 & 0 & 0 & 1 \\ 0 & 0 & 0 & 0 & 1 & 1 & 1 & 1 \\ 0 & 0 & 0 & 0 & 0 & 1 & 0 & 1 \\ 0 & 0 & 0 & 0 & 0 & 0 & 1 & 1 \\ 0 & 0 & 0 & 0 & 0 & 0 & 0 & 1 \end{bmatrix} \end{array}$$

We next note that G is self-inverse with respect to the field of integers mod 2, that is,

$$GG = I(\text{mod } 2) \tag{3.4}$$

This is most easily proved by induction. Denoting by G_N the resulting $2^N \times 2^N$ matrix for a given N, one may show that

$$G_{N+1} = \begin{pmatrix} G_N & G_N \\ 0 & G_N \end{pmatrix}$$

from which the induction follows directly.

Returning to (3.3), let V_f be the 1 X 2^N vector whose i-th entry is f evaluated at ordered state s_i. Then (3.2) may be extended to

$$V_f = \underset{\sim}{c}^T G \tag{3.5}$$

from which, using (3.4),

$$V_f G = \underset{\sim}{c}^T \tag{3.6}$$

that is, the coefficient vector $\underset{\sim}{c}$ may be obtained.

Returning to our earlier example with N = 3, what are the multinomials f_1, f_2, and f_3 guiding the elements 1, 2, and 3, respectively? From the state diagram we have

$$V_{f_1} = (0,1,0,1,0,0,1,1)$$
$$V_{f_2} = (0,0,0,0,1,1,0,0)$$
$$V_{f_3} = (1,1,0,0,0,1,1,0)$$

Thus

$$\underset{\sim}{c}_1^T = (0,1,0,0,0,1,1,0)$$
$$\underset{\sim}{c}_2^T = (0,0,0,0,1,0,1,0)$$
$$\underset{\sim}{c}_3^T = (1,0,1,0,1,1,0,0)$$

and

$$f_1 = x_1 + x_1x_3 + x_2x_3$$
$$f_2 = x_3 + x_2x_3$$
$$f_3 = 1 + x_2 + x_3 + x_1x_3$$

We can now define the function matrix F. F is the 2^N X 2^N matrix that has as its rows the coefficient vectors (c's) of the 2^N products of the N functions computed by the N elements of the switching net. That is, analogous to what was done in developing the g's, we create 2^N products from each of the 2^N subsets of the $f_1, f_2, \ldots, f_N$. Each such product will yield a multinomial function over our field. Corresponding to the empty set, we take the constant function 1. As before, the subsets may be sequenced in lexicographic order, thus ordering the rows of F.

In taking the products of these functions, it is convenient to observe that for any x, $x^p = x$, and that for any g, $2ag = 0$ for any integer a. Thus, for example, using f_1 and f_2 above, we discover that

$$\begin{aligned} f_1 f_2 &= (x_1 + x_1 x_3 + x_2 x_3)(x_3 + x_2 x_3) \\ &= x_1 x_3 + x_1 x_2 x_3 + x_1 x_3 + x_1 x_2 x_3 + x_2 x_3 + x_2 x_3 \\ &= 2x_1 x_3 + 2x_2 x_3 + 2x_1 x_2 x_3 = 0 \end{aligned}$$

In fact, the full F matrix for f_1, f_2, and f_3 above becomes

F =

	1	x_1	x_2	x_1x_2	x_3	x_1x_3	x_2x_3	$x_1x_2x_3$
1	1	0	0	0	0	0	0	0
f_1	0	1	0	0	0	1	1	0
f_2	0	0	0	0	1	0	1	0
f_1f_2	0	0	0	0	0	0	0	0
f_3	1	0	1	0	1	1	0	0
f_1f_3	0	1	0	1	0	1	1	0
f_2f_3	0	0	0	0	0	1	0	1
$f_1f_2f_3$	0	0	0	0	0	0	0	0

It is apparent that the matrix F is yet another equivalent representation of the net. In fact, the 3 X 8 matrix with rows corresponding to the functions f_1, f_2, and f_3 would suffice since we have shown that they are equivalent to the three Boolean transformations computed by the three net elements. The primary reason for augmenting the additional rows is to establish a relationship between F and T. That the function matrix and the transition matrix should be related does not seem surprising particularly since both are $2^N \times 2^N$.

The relationship involves the G matrix derived earlier. This matrix has as its columns representations of the states through the field of multinomials: that is, the i-th column of G corresponds to the i-th ordered state. Multiplying this matrix on the left by F results, for the i-th column, in the state to which the i-th ordered state goes. More precisely, if in F we look at the row corresponding to f_i, the Boolean transformation computed by net element i, then this row multiplied by column j of G is $\underset{\sim}{c}_i^T g$ evaluated at net state j, which yields the next state for element i. Considering all rows of F such multiplication yields the "next" net state to j under the foregoing representation.

Now recall that T is a matrix with exactly one 1 in each column such that the particular row in which it appears indicates the successor state. Consider multiplying G on the right by T. The j-th column of this product matrix will again be the state to which the j-th ordered state goes.

In other words, we have the relationship

$$FG = GT \tag{3.7}$$

In our example it may be verified that

$$FG = GT = \begin{bmatrix} 1 & 1 & 1 & 1 & 1 & 1 & 1 & 1 \\ 0 & 1 & 0 & 1 & 0 & 0 & 1 & 1 \\ 0 & 0 & 0 & 0 & 1 & 1 & 0 & 0 \\ 0 & 0 & 0 & 0 & 0 & 0 & 0 & 0 \\ 1 & 1 & 0 & 0 & 0 & 1 & 1 & 0 \\ 0 & 1 & 0 & 0 & 0 & 0 & 1 & 0 \\ 0 & 0 & 0 & 0 & 0 & 1 & 0 & 0 \\ 0 & 0 & 0 & 0 & 0 & 0 & 0 & 0 \end{bmatrix}$$

Since G is self inverse, (3.7) implies

$$F = GTG, \ T = GFG \tag{3.8}$$

This means we may easily obtain the functional representation from the transitional representation and vice versa.

In the preceding development, the reader may have failed to grasp the nature of the linearization of the net that has been effected through the F matrix. The next state function F given in (3.1) clearly need not be linear. However, employing the state representations provided by the columns of G, we see from (3.7) that the F matrix operates linearly (matrix multiplication) to provide the next net state. Moreover, as opposed to (3.2), the binary character of the states is preserved through G and the specific Boolean transformations computed by each element in the net are retained in F.

In summarizing this section we have developed four equivalent representations for a binary switching net. They are

1. A state diagram
2. A network (N,K) and an associated set of Boolean transformation
3. A transition matrix T
4. A function matrix F

It is obvious that representation (1) offers the most visual display of the net's behavior. However, in pursuing the theory of Chapter 4, representations (2), (3), and (4) must be used.

3.5 Applications

The basic structure and behavior of switching net models having been described, we now discuss several useful settings to which we may apply these models. After developing the details in Chapters 4 and 5, we return to each of these applications in Chapter 6 to extend our interpretation.

3.5.1 Genetic Control Systems

The pioneering work in applying binary switching net theory to genetic control systems has been done by Kauffman (1969, 1970, 1974). The net is interpreted as a cell's genetic system and each element as a gene. Several idealizations are incorporated to develop the analogy. First, time is presumed to occur in discrete clocked moments. The analytic advantages inherent in such a discrete scale offset the conceptual attractiveness of a continuous scale. Next, the pathway by which the output of a gene comes to influence another gene will be ignored. We will only acknowledge whether or not a direct connection exists. In this sense a network structure is well defined. Finally, each gene will be considered to be a binary switch capable only of being fully on or fully off. It is fair to ask whether genes tend to be able to assume finely graded levels of steady activity or whether they tend to be very active or very inactive. Kauffman offers several theoretical and experimental reasons to suggest the latter to be true. Hence Boolean transformations become appropriate descriptions of the response of a gene to its input information. The various net states correspond to differing states of gene activity for the cell. A regular pattern of gene activity or net states would correspond to a cycle. The collection of all cycles describes the various dynamic behaviors of the cell. With regard to connectance, gene control systems of cells are almost certainly not one-input systems. In known experimental cases, we always find at least two inputs. Moreover, behaviorally there are two obvious major disadvantages to a one input specification. First is that component failure could disconnect from the system all members of the hierarchy descendent from the defective gene. Second, one input switching nets do not provide the homeostatic tendency that gene systems exhibit. So gene control systems may be taken to have multiple inputs, often including feedback. The possibility of many inputs allows the possibility of redundancy and thus of more reliable behavior. Additionally the greater the number of inputs, the more subtle and complex the cell's behavior can be. However, with increased input connectance, the net will not show the restricted patterns of activity or the homeostatic behavior found in cells. To constrain the net model to more realistic behavior will require, as in the preceding application, the imposition of control through the selection of Boolean transformations governing the net elements. Formalization of equivalence classes of such constrained Boolean transformations is deferred to the next chapter. As with structural uncertainty, we deal with these classes via an ensemble approach. Attractive interpretations of these constraints are provided together with their ramifications in Chapter 6.

A second argument underscoring the need for control is the following. Current estimates of the number of genes in a higher metazoan cell range from 40,000 to 1,000,000. A metazoan with only 100,000 genes would then have $2^{100,000} \approx 10^{30,000}$ conceivable states of gene activity. At known rates of gene activity, a cell could not explore that dynamic space in billions of times the history of the universe. Although it is not known how small the subset of patterns of gene activity is, it must be quite small since one can recognize the same cell type over time and over cell divisions.

That a phenomenon involving 40,000 to 1,000,000 elements is complex is obvious. It is not reasonable to expect to discover virtually all control relations among such a large number of genes. Thus switching net models provide one approach to constructing an adequate picture of the architecture of cell control systems whose full details may never be directly known. Kauffman (1974, p. 167) offers a lucid statement of the philosophy behind this approach, capturing its essential "small scale properties to large scale behavior" idea. He asserts that

> [the] approach is to characterize any known small scale properties of the organization of cellular control systems, such as specifying the typical number of variables controlling any process and specifying the ways variations in the controlling process affect the controlled processes. Specification of such small scale local properties should be useful in two ways: (1) the local properties form the basis for hypothesis about the organization of larger control circuits; (2) the implications of the small scale properties for the large scale dynamic behavior of cellular control systems can be assessed. Systematic use of such local characteristics for both these purposes can be made by constructing a set of all the possible large control systems, each member of which is built using only these small scale properties.

(Note: We will return to this paragraph in Sec. 6.2, where we discuss Kauffman's modeling context at greater length.)

It is apparent that Kauffman is describing the development of an ensemble of systems to study. In his view the primary purpose in characterizing small scale properties and in constructing an ensemble of possible control systems is to examine the implications of known small scale features on large scale properties.

3.5.2 Management Strategy in Complex Organizations

The organizational analogy we develop interprets the switching net as a control system embedded in an organization. This analogy was first advanced in Walker and Gelfand (1979) and in Gelfand and Walker (1980). The net elements correspond to points in the organization where control information is used. The network structure represents the channels through which actual control information is passed in the system. The response of a control point to the possible contingencies presented to it by its sources of information is described by a Boolean transformation. Dichotomizing the input information as well as the elemental response restricts the scope of activities to which these switching net models may be applied. However, a considerable variety of processes are governed by yes-no, on-off, defective-nondefective, go-no go, and so on, decisions. Moreover, for most of the remainder, n-ary switching nets should provide an adequate model description. Although such nets are more complicated to examine, we are still dealing with essentially the same modeling perspective.

More specifically, if the activity is essentially a production process, the elements are seen as machines. The N machines represent N possibly different constituents, each of which is either being fabricated or not.

The cycle models the product, the transition states preceding the cycle model the startup period, and the entire set of cycles gives the inventory of products the particular control structure and particular set of constituents is capable of producing.

While the system size N is typically conceived of as quite large in much switching net modeling (at least order 10^4 or 10^5 in genetic applications), in the present instance N will probably be of order at most 10^2. But the relatively smaller system size does not preclude complexity. Even though an organizational control grid might be essentially fixed and in principle specifiable, the time required to know its details and thereby to make a prediction of system response might often exceed the time within which that prediction were needed. Therefore, the effective practice of management requires a body of knowledge concerning what behavior may be predicted under conditions of structural uncertainty. One method of accommodating that uncertainty is to study appropriate ensembles of structures. This method refocuses analytic interest on the behavioral properties of the structures so generated and, therefore, suggests a familiar idea, an ensemble approach to modeling.

The Boolean transformations are a key concept in this analogy. Management strategy, that is, the practice of management to achieve certain organizational behaviors, may be equated to imposing restrictions on the switching net model with regard to the set of available Boolean transformations. More precisely, the fact that an organization survives at all or perhaps in some cases prospers suggests that the completion of its individual processes does not often require excessive amounts of time. (The time frame is, of course, relative to the particular activity.) In other words, for the successful organization, repetitive activities must be completed in "reasonable" time epochs. Hence, if our nets are to be effective models of real-world organizations, they too must run through cycles having lengths which are empirically "reasonable." If N were, say, 100 and the net was uncontrolled, then cycle lengths might be on the order of $2^{100} \approx 10^{30}$. Under any plausible assumptions as to how fast the net moves from state to state, such a net could model no repetitive real world phenomenon. Therefore, as in the cell biology application, we must exercise control through the Boolean transformations governing the net elements to dramatically curtail cycle lengths. Perhaps surprisingly, the same equivalence classes developed for the genetic control application allow satisfying interpretations of these constraints as managerial strategy or policy options enhancing the usefulness of net models in this application. We offer these interpretations and their ramifications in Chapter 6.

3.5.3 Advertising Policy in Consumer Markets

The notion that, in many markets, consumer purchase behavior cannot be well articulated by static models is gaining increasing acceptance. Whether the product is inexpensive, such as soap or toothpaste, or costly, such as a car or hi-fi equipment, the buying process is typically exceedingly complex. Hopefully, some facets of the process can be usefully studied separately. One such facet is the word-of-mouth dynamics that influences prospective buyers in the choice of brand.

We can think of many markets in which it is abundantly clear that word-of-mouth is an important variable. Home buying is one example. In influencing a buyer's choice of brokerage firm, word-of-mouth has been cited (Hempel, 1969) as an important factor. The convenience and subtle influence of advice offered during the ordinary routine of social interaction ought not be underestimated. Effective marketing of a product mandates concern with these informal communication elements as well as the more familiar commercial channels. Advertising is to some extent predicated on the assumption that this informal facet of consumer behavior can be manipulated to create, for a particular brand, favorable word-of-mouth for that brand in the market. Hence successful advertising policy in employing available commercial channels must comprehend how these channels will affect word-of-mouth produced dispositions.

The worth of studying the phenomenon of word-of-mouth in certain dynamic systems is thus clear. That word-of-mouth is a complex phenomenon is apparent. It will typically involve a large number of individuals whose intercommunication structure is intricate and whose response to communication is nonhomogeneous. Happily, for our purpose binary switching net models are useful in studying the behavior of such systems.

The analogy for this application assumes that the net is a consumer group and that the net elements are households (customary units in marketing). For a given household, the associated influence structure is assumed to be relatively fixed. Hence we have a network. Our focus is on opinions favorable or unfavorable held by households with respect to a specific brand (retailer, service outlet, etc.) in the market. This dichotomy of opinion suggests that the reactions of an individual household to the opinion of other households with respect to the brand may be described via a Boolean function. Function inputs are then the household's influence sources, and the output value at any time represents the household's opinion on the brand at that time. The heterogeneity of households leads to a collection of Boolean transformations, one for each household, thus completing the switching net analogy. The network structure allows self-feedback, certainly, for many households, a component in molding opinion from influence sources.

The collective group opinion at a given time is represented by the state vector of the net at that time. One simple summary variable which may be monitored over time is the proportion of favorable opinions in the market. Another, a distance variable, may be developed to measure opinion constancy (i.e., brand loyalty over time). Note that cycles do not seem relevant in this application. In studying the behavior of these variables an ensemble approach naturally arises. Random net models are drawn according to a fixed group size and a fixed degree of influence upon a household. These replications enable one to observe how the average and variability of these process variables change with time.

Advertising efforts are presumed to be directed at how households transact their word-of-mouth opinions rather than being directed at modifying the existing transaction patterns. Again the key role of the Boolean transformation is revealed. Structural control through the selection of these mappings to achieve desirable behavior of the system variables may be viewed as adopting different advertising policies. As in the previous examples, the types of control we consider have appealing interpretations. In the present case they emerge as distinctive advertising styles (see Chapter 6).

References

Alben, R. and Boutron, P. (1975). Continuous random network model for amorphous solid water. *Science, 187*, 430-432.

Arbib, M. A. (1972). *The Metaphorical Brain*. New York: Wiley.

Atlan, H., Fogelman-Soulie, F., Salomon, J., and Weisbuch, G. (1982). Random Boolean networks. *Journal of Cybernetics*, to appear.

Babcock, A. K. (1976). Logical probability models and representation theorems on the stable dynamics of the genetic net. Doctoral dissertation, State University of New York at Buffalo.

Bell, A. J. and Dean, P. (1972). Structure of vitreous silica. Validity of random network theory. *Philosophical Magazine*, *25*, 1381-1398.

Blumenson, L. E. (1970). Random walk and the spread of cancer. *Journal of Theoretical Biology*, *27*, 273-290.

Caianiello, E. R. (1973). Some remarks on the temporal linearization of general and linearly separable Boolean functions. *Kybernetik*, *12*, 90-93.

Cull, P. (1971). Linear analysis of switching nets. *Kybernetik*, *8*, 31-39.

Doreian, P. (1974). On the connectivity of social networks. *Journal of Mathematical Sociology*, *3*, 245-258.

Gelfand, A. E. and Walker, C. C. (1980). A system theoretic approach to the management of complex organizations: Management by consensus level and its interaction with other management strategies. *Behavioral Science*, *15*, 250-60.

Hempel, D. (1969). The role of the real estate broker in the home buying process. CREUES Real Estate Report 7. Center for Real Estate and Urban Economic Studies, University of Connecticut, Storrs.

Kauffman, S. (1969). Metabolic stability and epigenesis in randomly constructed genetic nets. *Journal of Theoretical Biology*, *22*, 437-467.

Kauffman, S. (1970). The organization of cellular genetic control systems. *Mathematics in the Life Sciences*, *3*, 63-116.

Kauffman, S. (1974). The large scale structure and dynamics of general control circuits: An ensemble approach. *Journal of Theoretical Biology*, *44*, 167-190.

Latour, P. L. (1963). The neuron as a synchronous unit, *Progress In Brain Research*, Vol. 2: *Nerve, Brain and Memory Models*. Amsterdam: Elsevier, pp. 30-36.

McCulloch, W. S. and Pitts, W. H. (1943). A logical calculus of the ideas immanent in nervous activity. *Bulletin of Mathematical Biophysics, 5,* 115-133.

Rapoport, A. (1948). Cycle distributions in random nets. *Bulletin of Mathematical Biophysics, 10,* 145-157.

Rapoport, A. (1954). Random nets with transitivity bias. Proceedings of the Symposium on Information Networks, Brooklyn, N.Y., pp. 187-197.

Rashevsky, N. (1960). *Mathematical Biophysics,* New York: Dover, pp. 230-271.

Sherlock, R. A. (1979a). Analysis of the behavior of Kauffman binary network I: State space description and the distribution of limit cycle lengths. *Bulletin of Mathematical Biology, 14,* 687-705.

Sherlock, R. A. (1979b). Analysis of the behavior of Kauffman binary networks II: The state cycle fraction for networks of different connectivities. *Bulletin of Mathematical Biology, 41,* 707-724.

Shimbel, A. (1948). An analysis of theoretical systems of differentiating nervous tissue. *Bulletin of Mathematical Biophysics, 10,* 131-143.

Shimbel, A. (1951). A note on the problem of cycles in random nets. *Bulletin of Mathematical Biophysics, 13,* 319-321.

Solomonoff, R. (1952) An exact method for the computation of the connectivity of random nets. *Bulletin of Mathematical Biophysics, 14,* 153-157.

Solomonoff, R. and Rapoport, A. (1951). Connectivity of random nets. *Bulletin of Mathematical Biophysics, 13,* 131-143.

Stubbs, D. F. (1977). Connectivity and the brain. *Kybernetics, 6,* 1-7.

Stubbs, D. F. and Good, P. I. (1976). Connectivity in random networks. *Bulletin of Mathematical Biology, 38,* 295-304.

Walker, C. C. and Gelfand, A. E. (1979). A system theoretic approach to the management of complex organizations: Management by exception, priority and input span in a class of fixed structure models. *Behavioral Science, 24,* 112-120.

4

Theoretical Results on the Behavior of Switching Net Models

4.1 Introduction

How do switching nets behave? We have introduced several behavioral variables to try to provide a descriptive answer. In terms of these variables, a first dichotomy splits the question into analytical and empirical investigation. It is the intent of this chapter to consider theoretical results. Discussion of the considerable body of simulation work will be deferred to Chapter 5. This empirical evidence is most important in enabling one to "get a feel" for switching net behavior. Additionally, these simulations are usually developed from ensemble-based notions such as those we have previously discussed. However, some theoretical preparation will hopefully add greater insight and significance to this work. Sections 4.2 to 4.6 are intended to provide the necessary background. Moreover, analytical limitations have tended to hinder the impact of the type of modeling enterprise we are offering in this book. These sections are intended to help the reader to appreciate these limitations as well as to see what has been done. Proofs are supplied for most results, although they will hold little interest for many readers and may be skipped with little loss of comprehension. The reader interested in such theoretical research may benefit from seeing the diversity of techniques employed.

4.2 Relationships Between the Transition Matrix, the Function Matrix, and the Cycle Space

Recall the $2^N \times 2^N$ transition matrix T defined in Sec. 3.4; the entries in T, T_{ij}, are defined by

$$T_{ij} = \begin{cases} 1 & \text{if state i is the successor to state j} \\ 0 & \text{otherwise} \end{cases}$$

Hence T has exactly one 1 per column. Moreover, $T_{ii} = 1$ iff state i is on a cycle of length 1. Thus $Tr(T)$ [$Tr(T) \equiv$ trace of $T = \Sigma t_{ii}$] gives the number of elements on cycles of length 1. Extending this notion it is apparent that

$$Tr(T^m) = \text{number of states on cycles whose length divides m} \tag{4.1}$$

and that

$$Tr(T^{2^{N!}}) = \text{number of cyclic states}$$

$$2^N - Tr(T^{2^{N!}}) = \text{number of transient states}$$

$$Tr(T^{r!}) = \text{number of states on cycles of length at most r}$$

We note that (4.1) is less than satisfactory in the sense that we would prefer a matrix A^m obtainable from T such that

$$Tr(A_m) = \text{number of states on cycles whose length is exactly m} \tag{4.2}$$

Let C_m = {primes $\leq$ m which appear in the prime representation of m}, i.e., with power $\leq$ 1, and let Γ_m = number of elements in C_m. The number of subsets of Γ_m is 2^{Γ_m} and the number of subsets of size k is $\binom{\Gamma_m}{k} \equiv \Gamma_{m,k}$. At a given k, let j index the subsets of size k so that 2^{Γ_m} subsets may be denoted by C_{kj}, $k = 0, \ldots, m$, $j = 1, \ldots, \Gamma_{m,k}$. Let g_{kj} equal m divided by the product of all the elements in C_{kj}. For the empty set we take $g_{01} = m$. Then we have the following result.

Theorem 1. For each m, $m = 1, \ldots, 2^N$, let

$$A_m = \sum_{k=0}^{\Gamma_m}(-1)^k \sum_{j=1}^{\Gamma_{m,k}} T^{g_{kj}}$$

Then $\mathrm{Tr}(A_m)$ = number of states on cycles whose length is exactly m.

Proof. The most direct proof employs a straightforward but tedious inclusion-exclusion argument.

We illustrate with an example. Suppose that we seek the number of states on cycles of length 12 (i.e., $N \geq 4$). Then

$$m = 12, \quad C_{12} = \{2,3\}, \quad \Gamma_{12} = 2$$

$$\Gamma_{12,0} = 1, \quad \Gamma_{12,1} = 2, \quad \Gamma_{12,2} = 1$$

$$g_{01} = 12, \quad g_{11} = 6, \quad g_{12} = 4, \quad g_{22} = 2$$

and

$$A_{12} = T^{12} - T^6 - T^4 + T^2$$

We now develop ideas due to Cull (1971) relating the cycle space to the characteristic equation of T. Suppose that T results in a cycle space having m_0 transient elements and m cycles of lengths $r_1, r_2, \ldots, r_m$, respectively. Consider the characteristic polynomial of T, $|T - \lambda I|$, where operations are performed in the real field. We argue that this polynomial will have the form

$$\pm\lambda^{m_0} \prod_{i=1}^{m} (\lambda^{r_i} - 1) \tag{4.3}$$

Rearrange the rows and columns of T so that, proceeding from left to right (or top to bottom), we first encounter all the "first" states (i.e., states into which no state maps). Rows corresponding to first states in $T - \lambda I$ will have $-\lambda$ on the diagonal as the only nonzero entry. Expanding the determinant of $T - \lambda I$ by these rows produces $-\lambda$ to a power equal to the number of first states multiplied by a reduced determinant. Now consider each of the states that can be reached only from eliminated states. In the reduced determinant, rows corresponding to each of these states will again have $-\lambda$ on the diagonal as the only nonzero entry. Expanding in terms of these rows and continuing the process, it is clear that we may eliminate all of the transient states, resulting in a factor of $(-\lambda)^{m_0}$ multiplied by a reduced determinant involving only cyclic states. For this reduced determinant again rearrange rows and columns so that the various cycles form blocks within the matrix. Then the reduced determinant will be the product of the determinants of each of the blocks. For a cycle of length r the determinant will be $\pm(\lambda^r - 1)$; that is, all characteristic roots are r-th roots of 1. Hence (4.3) follows.

The binary characteristic polynomial of T (i.e., $|T + \lambda I|$, where operations are performed in the binary field) is, from the above, seen to be of the form

$$\lambda^{m_0} \prod_{i=1}^{m} (\lambda^{r_i} + 1) \tag{4.4}$$

In the *binary* form using (3.4) and (3.8),

$$\begin{aligned} |F + \lambda I| &= |GTG + \lambda GG| \\ &= |G|\ |T + \lambda I|\ |G| \\ &= |GG|\ |T + \lambda I| \\ &= |T + \lambda I| \end{aligned}$$

that is, F and T have the same *binary* characteristic polynomial. Unfortunately, we do not always obtain a unique factorization of these binary polynomials into the form (4.4) since

$$\lambda^{2^j} + 1 = (\lambda + 1)^{2^j}$$

for example, a cycle of length 2 may appear to be two cycles of length 1.

From the Cayley-Hamilton theorem, F satisfies its characteristic equation; that is,

$$F^m \prod_{i=1}^{m_0} (F^{r_i} + I) = 0 \tag{4.5}$$

and, furthermore, any polynomial of lower degree that F satisfies must divide the characteristic polynomial. In fact, there exists a minimal equation or equation of lowest degree which F satisfies. In this equation, which must resemble (4.5), the power of the F term will be the length of the longest chain of transient states, while the powers of F in the product will be distinct cycle lengths and such that any other cycle lengths divide one of these powers.

To illustrate, consider the state diagram and associated T matrix presented above (3.1). Rearranging rows and columns, we obtain

	111	110	100	101	011	000	001	010
111	0	0	0	0	0	0	0	0
110	0	0	0	0	0	0	0	0
100	1	1	0	0	0	0	0	0
101	0	0	1	0	1	0	0	0
011	0	0	0	1	0	0	0	0
000	0	0	0	0	0	0	0	1
001	0	0	0	0	0	1	0	0
010	0	0	0	0	0	0	1	0

It is clear that

$$|T - \lambda I| = (-\lambda)^3(1 - \lambda^3)(1 - \lambda^2)$$

indicating three transient states, a cycle of length 3, and a cycle of length 2. In the binary field

$$|F + \lambda I| = \lambda^3(\lambda^3 + 1)(\lambda^2 + 1)$$

and the characteristic equation is

$$F^3(F^3 + 1)(F^2 + 1) = 0 \qquad (4.6)$$

The reader may establish that

$$F(F^3 + 1)(F^2 + 1) \neq 0 \qquad (4.7)$$

but that

$$F^2(F^3 + 1)(F^2 + 1) = 0 \qquad (4.8)$$

Hence (4.8) is the minimal equation of F and reveals that the longest transient chain is of length 2.

Next we consider the eigenvectors of T or, equivalently in the binary field, of F. As noted above (3.4), the matrix G has as its columns representations of the 2^N state vectors through the field of multinomials. Using these representations, we have shown via (3.8) and associated remarks the linearization

$$X^{t+1} = FX^t$$

where X^{t+1} is the successor state to X^t. Consider any cycle of F consisting of states, say, $X_{\alpha_1}, \ldots, X_{\alpha_r}$ (i.e., r columns of G). By (4.9) it is clear that

$$F \sum_{i=1}^{r} X_{\alpha_i} = \sum_{i=1}^{r} X_{\alpha_i}$$

That is, the sum of the vectors corresponding to states on a particular cycle is an eigenvector for the characteristic root 1. Moreover, if $X_{\beta_1}, \ldots, X_{\beta_s}$ are the states on another cycle in F, we have

$$F\left(\sum_{i=1}^{r} X_{\alpha_i} + \sum_{j=1}^{s} X_{\beta_j}\right) = \sum_{i=1}^{r} X_{\alpha_i} + \sum_{j=1}^{s} X_{\beta_j}$$

That is, the sum of the states on any set of cycles is also an eigenvector for the characteristic root 1.

Conversely, consider any eigenvector Z corresponding to the characteristic root 1, that is,

$$FZ = Z \tag{4.10}$$

Since the columns of G form a basis for the 2^N-dimensional vector space,

$$Z = Ga = \sum_{k=1}^{2^N} a_k X_k$$

where the X_k are the column vectors of G. In fact, by (3) we may immediately solve for a:

$$a = GZ$$

The vector a has only 0's and 1's as components. If $a_i = 1$, state X_i is in Z's representation. From (4.10)

$$F^{2^N} Z = Z$$

whence if state X_i is transient, $a_i = 0$ in Z's representation. That is, after 2^N transitions no state (i.e., column of G) can be in a transient state. Hence for X_k corresponding to a transient state, $F^{2^N} X_k \neq X_k$ (i.e., $a_k = 0$). Therefore, only cyclic states are included in Z's representation. If state X_i is such a state and state X_j is on the same cycle as X_i, then state X_j is included in Z's representation. This follows since $X_i = F^m X_j$ for some $m \geq 1$, implying that

$$Z = F^m Z = \sum_{k=1}^{2^N} a_k F^m X_k$$

In this sum, at $k = j$ we have $a_j F^m X_j = a_j X_i$. But state X_i is in Z's representation, so that $a_j = 1$ and therefore state j is in Z's representation. Continuing this argument, if Z's representation contains any state on a cycle, it contains all states on that cycle. Hence the states in the representation of Z will be a set of cycles. In terms of the transition matrix, we may write (4.10) as

$$GTGZ = Z$$

or

$$TGZ = GZ$$

that is,

$$Ta = a \tag{4.11}$$

Note the parallel between (4.10) and (4.11). Also note that the length of a is the number of cyclic states in Z's representation.

Continuing our illustrative example, suppose that

$$Z^T = (1,1,0,0,1,1,1,0)$$

We may immediately verify that $FZ = Z$ and that $a^T = (1,0,1,0,1,1,1,1)$. Hence we see that all five cyclic states of our example (101, 001, 000, 001, and 010, corresponding to columns 6, 7, 1, 5, and 3 of G) are included in Z's representation.

By contrast, the zero eigenvectors corresponding to the characteristic root 0 give information about the transient chains of states. Suppose that

$$FZ = 0 \tag{4.12}$$

Then if $X_1, \ldots, X_m$ are the states in Z's representation, we have

$$F(X_1 + \ldots + X_m) = 0$$

This equation shows that for F every eigenvector corresponding to the root 0 is made up of an even number of states which can be broken down pairwise such that the two members of each pair go to the same next state. In terms of T, (4.12) becomes

$$GTGZ = 0$$

or

$$TGZ = 0 \tag{4.13}$$

that is,

$$Ta = 0.$$

4.3 Random Nets

In this section we discuss material from Gelfand (1982).

To obtain behavioral benchmarks and to demonstrate theoretically the need for exercising control, we first consider completely uncontrolled switching nets. Such a concept suggests that we wish to think of there being no restriction on the interconnectance structure between elements and no restriction on the selection of Boolean transformation at each net element. At each element this means that the response of an element to its input information is as likely to be 0 as it is to be 1. Otherwise, some control, favoring one or the other output value, is present. Moreover, the response of an element in no way affects (controls) the response of any other element. These local assumptions assert that the individual elemental responses may be viewed as observations on a collection of 2^N independent and identically distributed Bernoulli ($p = \frac{1}{2}$) variables. Hence no matter what state the net is presently in, its next state is as likely to be any of the 2^N possible states, that is,

$$P(\text{next net state is } X_j \text{ given current net state is } X_i) = \tfrac{1}{2}^N \quad j = 1, 2, \ldots, 2^N.$$

Suppose that we construct such a net. That is, state by state we make a random (equally likely) assignment of a successor state. Having made these assignments, we now have a well-defined net [one of $(2^N)^{2^N}$ equally likely possibilities]. In principle, we could readily draw its state diagram or cycle space. From Chapter 3's work we could obtain its network representation and associated Boolean transformations, its transition matrix, and its function matrix. A net developed in this manner might be called a *completely* random net rather than a completely uncontrolled net. In fact, in the literature such a net is usually called a random net. This circumstance is unfortunate since it is potentially misleading. There are an infinity of "random" mechanisms (probability distributions) which can be used in making next-state assignments. Indeed, exercising control over a net corresponds to selecting a random mechanism other than "equally likely," but randomness is still present. Nonetheless, we abide with convention.

An alternative development of a random net arises through the transition matrix. There are $(2^N)^{2^N}$ possible transition matrices; that is, column by column we may assign a 1 to any one of the 2^N rows. Suppose that we make an equally likely selection of a transition matrix. Such a procedure clearly creates a random net in a fashion equivalent to the procedure described above. This equivalence of the state diagram and transition matrix approaches is a special case of the decision theoretic equivalence, for finite state spaces, of behavioral decision rules and randomized decision rules, respectively (see Ferguson, 1967, pp. 24-25).

Once the equivalence between random nets and random transition matrices has been identified, we may immediately simplify the entire problem. Let $X = \{x_1, \ldots, x_n\}$ be a finite set of n elements and let $\mathcal{T}$ be the set of all transformations from X into itself. The equally likely selection of a random transformation on a finite set of n elements is now seen to include the selection of a random net with N elements. All we would require is that $n = 2^N$ and that the x_i be the binary state vectors for the net. This description suffices for examination of the cycle structure of a net but sacrifices the net's binary character. For a completely random net the binary aspect is immaterial. In the interest of greater generality and notational convenience, for the remainder of this section we examine the cycle space of a random transformation on a finite set.

The extant literature in this area is limited. Early work by Gontcharoff (1944) considers the distribution of cycles in permutations of a finite number of elements. Riordan (1958) updates this discussion in his Chapter 4. Rubin and Sitgreaves (1954), in a very long and detailed article, consider some aspects of the cycle space without formally asserting so. Harris (1960) extends their work and includes some results discussed here but obtained from a different point of view. Katz (1955) and his student, Folkert (1955), examine the expected number of cycles. Cull (1978) studies the problem in a system setting using binary switching nets (although to no particular advantage) and develops some approximate results on the expected number of cycles and cyclic states.

Consider now the selection of a random (equally likely) transformation T from $\mathcal{T}$. This selection is conveniently accomplished as a sequence of n independent multinomial trials where the j-th trial chooses the successor to state j in an equiprobable fashion from among the n elements in X. This approach clearly results in an equiprobable selection of the n^n elements in *T*.

Then Tr(T), the number of states on cycles of length 1, is obviously distributed as binomial (n,1/n) with expectation, E(Tr(T)) = 1, and variance, var(Tr(T)) = (n - 1)/n. The probability that T has no cycles of length 1 is $((n - 1)/n)^n$; the probability that state i is a successor state is $1 - ((n - 1)/n)^n$. As $n \to \infty$ these probabilities tend to e^{-1} and $1 - e^{-1}$, respectively. The limiting distribution (as $n \to \infty$) of Tr(T) is Poisson(1).

More generally, we pose the following questions regarding the cycle space of a random transformation:

1. What is the probability that state x_i is on a cycle of length r?
2. What is the joint probability that state x_i is on a cycle of length r and that state x_j is on a cycle of length s?
3. What is the expected number of cycles of length r and the expected number of states on cycles of length r?
4. What is the distribution of the number of cycles of length r and of the number of states on cycles of length r?
5. What is the joint distribution of the number of cycles of length r and the number of cycles of length s? Of the number of states on cycles of length r and the number of states on cycles of length s?
6. What is the expected number of cycles and the expected number of states on cycles?

7. What is the distribution of the number of cycles and of the number of states on cycles?
8. What is the expected length of a cycle?

In what follows we provide exact or asymptotic answers to all of these questions. As a first step an approach using a sequence of square arrays can be employed advantageously to study moments of the cycle structure variables.

4.3.1 Moments of Cycle Structure Variables

For a set of X of n elements and T selected at random from *T*, consider the n x n array of random variables.

$$\begin{matrix} D^n_{11} & \cdots & D^n_{1n} \\ D^n_{21} & \cdots & D^n_{2n} \\ \vdots & & \vdots \\ D^n_{n1} & \cdots & D^n_{nn} \end{matrix} \tag{4.14}$$

where

$$D^n_{ri} = \begin{cases} 1 & \text{if state } x_i \text{ is on a cycle of length } r \\ 0 & \text{otherwise} \end{cases} \tag{4.15}$$

From this array we are interested in the following variables.

$$S_{n,r} = \sum_{i=1}^{n} D^n_{ri} = \text{number of states on a cycle of length } r \tag{4.16}$$

$$T_{n,r} = S_{n,r}/r = \text{number of cycles of length } r \tag{4.17}$$

$$C^n_i = \sum_{r=1}^{n} D^n_{ri} = \begin{cases} 1 & \text{if state } x_i \text{ is on a cycle} \\ 0 & \text{otherwise} \end{cases} \tag{4.18}$$

$$U_n = \sum_{r=1}^{n} S_{n,r} = \sum_{i=1}^{n} C_i^n = \text{number of states on cycles} \tag{4.19}$$

$$V_n = \sum_{r=1}^{n} T_{n,r} = \text{number of cycles} \tag{4.20}$$

Note that while a row sum ($S_{n,r}$) may exceed 1, by definition the column sums (C_i^n) are still 0-1 random variables. In fact, $P(C_i^n = 0)$ is the probability that state i is transient.

For any fixed r the joint distribution of $D_{r1}^n, \ldots, D_{rn}^n$ of any subset will be that of a collection of dependent interchangeable random variables. The marginal distribution of any D_{ri}^n is given by

$$\begin{aligned} P(D_{ri}^n = 1) &= P\binom{\text{state } x_i \text{ is on a cycle}}{\text{of length } r} \\ &= \binom{n-1}{r-1}(r-1)!\,\frac{1}{n^r} = \frac{1}{n}\,\frac{(n)_r}{n^r} \end{aligned} \tag{4.21}$$

where $(n)_r$ is the falling factorial of r terms starting at n. We thus have $E(D_{ri}^n)$ and $\text{var}(D_{ri}^n)$ and note that as $n \to \infty$, both tend to 0.

We can immediately obtain the expectation for each of the variables in (4.16) through (4.20), that is,

$$E(S_{n,r}) = (n)_r/n^r \to 1 \text{ as } n \to \infty \tag{4.22}$$

$$E(T_{n,r}) = \frac{1}{r}(n)_r/n^r \to \frac{1}{r} \text{ as } n \to \infty \tag{4.23}$$

$$E(C_i^n) = \frac{1}{n}\sum_{r=1}^{n}(n)_r/n^r \to 0 \text{ as } n \to \infty \tag{4.24}$$

$$E(U_n) = \sum_{r=1}^{n}(n)_r/n^r \to \infty \text{ as } n \to \infty \tag{4.25}$$

$$E(V_n) = \sum_{r=1}^{n} \frac{1}{r}(n)_r/n^r \to \infty \text{ as } n \to \infty \tag{4.26}$$

The limits mean that as the number of states increases, the probability of any particular state being on a cycle tends to 0, but the expected number of cyclic states and expected number of cycles tends to ∞.

To obtain moments of order 2 for the variables above, we require the joint distribution of any pair, D^n_{ri}, D^n_{sj}. We have three cases: (1) $r \neq s$, $i \neq j$, (2) $r = s$, $i \neq j$, and (3) $r \neq s$, $i = j$.

For case 1 we have

$$P(D^n_{ri} = 1, D^n_{sj} = 1) = \begin{cases} \dfrac{1}{n(n-1)} \dfrac{(n)_{r+s}}{n^{r+s}}, & r + s \leq n \\ 0, & r + s > n \end{cases} \tag{4.27}$$

For case 2

$$P(D^n_{ri} = 1, D_{rj} = 1) = \begin{cases} \dfrac{r-1}{n(n-1)} \dfrac{(n)_r}{n^r} + \dfrac{1}{n(n-1)} \dfrac{(n)_{2r}}{n^{2r}}, & 2r \leq n \\ \dfrac{r-1}{n(n-1)} \dfrac{(n)_r}{n^r}, & 2n \geq 2r > n \\ 0, & r > 2n \end{cases} \tag{4.28}$$

For case 3 we have two exclusive events so that

$$P(D^n_{ri} = 1, D^n_{si} = 1) = 0 \tag{4.29}$$

In each case using (4.21) we may obtain expressions for the three remaining joint events.

Continuing, we have in case 1

$$\operatorname{cov}(D_{ri}^n, D_{sj}^n) = \begin{cases} \frac{1}{n(n-1)} \frac{(n)_{r+s}}{n^{r+s}} - \frac{1}{n^2} \frac{(n)_r (n)_s}{n^{r+s}}, & r + s \leq n \\ -\frac{1}{n^2} \frac{(n)_r (n)_s}{n^{r+s}}, & r + s > n \end{cases} \quad (4.30)$$

in case 2

$$\operatorname{cov}(D_{ri}^n, D_{rj}^n) = \begin{cases} \frac{r-1}{n(n-1)} \frac{(n)_r}{n^r} + \frac{1}{n(n-1)} \frac{(n)_{2r}}{2r} - \frac{1}{n^2} \frac{[(n)_r]^2}{n^{2r}}, & 2r \leq n \\ \frac{r-1}{n(n-1)} \frac{(n)_r}{n^r} - \frac{1}{n^2} \frac{[(n)_r]^2}{n^2}, & 2r > n \end{cases} \quad (4.31)$$

and in case 3

$$\operatorname{cov}(D_{ri}^n, D_{si}^n) = -\frac{1}{n^2} \frac{(n)_r (n)_s}{n^{r+s}} \quad (4.32)$$

In all cases these covariances tend to 0 as $n \to \infty$, a fact that can be inferred without computation from the Cauchy-Schwarz inequality, that is, because $\operatorname{var}(D_{ri}^n) \to 0$ as $n \to \infty$. Continuing, we have

$$\operatorname{cov}(S_{n,r}, S_{n,s}) = \begin{cases} \frac{(n)_{r+s}}{n^{r+s}} - \frac{(n)_r (n)_s}{n^{r+s}}, & r \neq s,\ r + s \leq n \\ -\frac{(n)_r (n)_s}{n^{r+s}}, & r \neq s,\ r + s > n \end{cases} \quad (4.33)$$

$$\operatorname{cov}(T_{n,r}, T_{n,s}) = \frac{1}{rs} \operatorname{cov}(S_{n,r}, S_{n,s}) \quad (4.34)$$

$$\operatorname{var}(S_{n,r}) = \begin{cases} r\frac{(n)_r}{n^r} + \frac{(n)_{2r}}{n^{2r}} - \frac{[(n)_r]^2}{n^{2r}}, & 2r \leq n \\ r\frac{(n)_r}{n^r} - \frac{[(n)_r]^2}{n^{2r}}, & 2r > n \end{cases} \quad (4.35)$$

$$\mathrm{var}(T_{n,r}) = \frac{1}{r^2}\,\mathrm{var}(S_{n,r}) \tag{4.36}$$

$$\mathrm{cov}(C_i^n, C_j^n) = \frac{2}{n(n-1)} \sum_{r=1}^{n} (r-1)\,\frac{(n)_r}{n^r} - \frac{1}{n^2}\left[\sum_{r=1}^{n} \frac{(n)_r}{n^r}\right]^2 \tag{4.37}$$

$$\mathrm{var}(C_i^n) = \frac{1}{n}\sum_{r=1}^{n} \frac{(n)_r}{n^r} - \frac{1}{n^r}\left[\sum \frac{(n)_r}{n^r}\right]^2 \tag{4.38}$$

$$\mathrm{var}(U_n) = \sum_{r=1}^{n} (2r-1)\frac{(n)_r}{n^r} + \left[\sum_{r=1}^{n} \frac{(n)_r}{n^r}\right]^2 \tag{4.39}$$

$$\mathrm{var}(V_n) = \sum_{r=1}^{n} \frac{1}{r}\frac{(n)_r}{n^r} - \left[\sum_{r=1}^{n} \frac{1}{r}\frac{(n)_r}{n^r}\right] + \sum_{\substack{r,s>1 \\ r+s\le 1}} \frac{1}{rs}\,\frac{(n)_{r+s}}{n^{r+s}} \tag{4.40}$$

From these expressions it is clear that $S_{n,r}$ and $S_{n,s}$ (also $T_{n,r}$ and $T_{n,s}$) are always negatively correlated but asymptotically uncorrelated. Also, $\lim_{n\to\infty} \mathrm{var}(S_{n,r}) = r$, $\lim_{n\to\infty} \mathrm{var}(T_{n,r}) = 1/r$. Since $\lim_{n\to\infty} \mathrm{var}(C_i^n) = 0$, we have $\lim_{n\to\infty} \mathrm{cov}(C_i^n, C_j^n) = 0$. Finally, $\mathrm{var}(U_n)$ and $\mathrm{var}(V_n)$ both tend to ∞ as $n \to \infty$, although this is most easily seen from results in Sec. 4.3.3.

To obtain moments of order m for the variables above, consider any subset of size m of the D_{ri}^n. Suppose first that all m variables are in the same row of

(4.14). Taking $mr \leq n$ and recognizing the exchangeability of the variables, we seek

$$P_{n,m,r} \equiv P(\text{states } x_{\alpha_1}, x_{\alpha_2}, \ldots, x_{\alpha_m} \text{ are each on a cycle of length } r) \tag{4.41}$$

$$= P(D^n_{r\alpha_1} = D^n_{r\alpha_2} = \cdots = D^n_{r\alpha_m} = 1)$$

To obtain an expression for this probability, consider all possible partitions of m with no part greater than r. If a given partition has parts $m_1, \ldots, m_j$, let $n(m_1, \ldots, m_j)$ be the number of ways to allocate m distinct objects into j like cells with m_i in cell i $(\sum_{i=1}^{j} m_i = m)$. Also associate with $m_1, m_2, \ldots, m_j$ the event $A_{n,r}(m_1, \ldots, m_j)$ defined by {states $x_{\alpha_1}, \ldots, x_{\alpha_{m_1}}$ on the same cycle of length r, etc.}. If S_m is the set of all partitions of m and $S_{m,r}$ is the set of all partitions of m with no part greater than r, then

$$P_{n,m,r} = \sum_{S_{m,r}} n(m_1, \ldots, m_j) P[A_{nr}(m_1, m_2, \ldots, m_j)] \tag{4.42}$$

where

$$P\left[A_{nr}(m_1, m_2, \ldots, m_j)\right] = \frac{1}{(n)_m} (n)_{jr} n^{-jr} \left[(r-1)!\right]^j \left[\prod_{i=1}^{j} (r-m_i)!\right]^{-1} \tag{4.43}$$

Using (4.41) with appropriate subsets of size m - 1, we may obtain the complete joint distribution of the m $D^n_{r\alpha_i}$. If, on the other hand, the m D^n_{ri} are all in the same column of (4.14), say $D^n_{\alpha_1 i}, \ldots, D^n_{\alpha i_m}$, in accordance with (4.21) their joint distribution will be multinomial with associated

$$P_{\alpha_j} = \frac{1}{n} \frac{(n)_{\alpha_j}}{n^{\alpha_j}}, \; j = 1, \ldots, m$$

Extending the ideas above, we may obtain the joint distribution of any subset of size m of D^n_{ri}. We omit the details.

4.3.2 Exact Distributions

Returning to the variables in (4.16) to (4.20), we have noted that C^n_i is a 0-1 variable with success probability given by (4.24). We can obtain the exact distributions of U_n following ideas given by Rubin and Sitgreaves (1954). Given T, for any $x \in X$, define the set of all successors to x, S(x), that is,

$$S(x) = \{x^1: \; T^r x = x^1 \text{ for some } r \geq 0\}$$

By definition $x \in S(x)$ and S(x) includes all the cyclic states associated with x (although x is, of course, not necessarily cyclic). Then with $k \geq r + 1$,

P(x has k successors, S(x) has cycle of length r, x is not cyclic) $= P(Tx \neq x; T^2x \neq Tx, T^2x \neq x; T^3x \neq T^2x, T^3x \neq Tx, T^3x \neq x; T^{k-1}x \neq T^{k-2}x, \ldots, T^{k-1}x \neq x; T^kx = T^{k-r}x)$

$$= \frac{n-1}{n} \cdot \frac{n-2}{n} \cdot \ldots \cdot \frac{n-(k-1)}{n} \cdot \frac{1}{n}$$

$$= \frac{(n)_k}{n^{k+1}}$$

Thus

P[S(x) has cycle of length r, x is not cyclic]

$$= \sum_{k=r+1}^{n} \frac{(n)_k}{n^{k+1}} \tag{4.44}$$

But

P[S(x) has cycle of length r, x is not cyclic]

$$= \sum_{u=r}^{n} \text{P[S(x) has cycle length of r, x is not cyclic, } U_n = u]$$

$$= \sum_{u=r}^{n} \text{P[S(x) has cycle of length r given x is not cyclic, } U_n = u] \cdot \text{P(x is not cyclic given } U_n = u) \cdot P(U_n = u)$$

$$= \sum_{u=r}^{n} \left[1 \cdot \frac{u-1}{u} \cdot \frac{u-2}{u-1} \cdot \ldots \cdot \frac{u-(r-1)}{u-(r-2)} \cdot \frac{1}{u-(r-1)}\right] \cdot \frac{n-u}{n} \cdot P(U_n = u)$$

$$= \sum_{u=r}^{n} \frac{n-u}{nu} P(U_n = u) \qquad (4.45)$$

But (4.44) and (4.45) are equal for all r, implying that

$$\sum_{k=r+1}^{n} \frac{(n)_k}{n^{k+1}} - \sum_{k=r+2}^{n} \frac{(n)_k}{n^{k+1}} = \sum_{u=r}^{n} \frac{n-u}{nu} P(U_n = u) - \sum_{u=r+1}^{n} \frac{n-u}{nu} P(U_n = u)$$

from which

$$P(U_n = u) = \frac{(n)_u}{n^{u+1}} u, \quad u = 1, 2, \ldots, n \qquad (4.46)$$

From (4.46), $P(U = n) = n!/n^n$. This is seen directly by noting that $U_n = n$ iff T is 1-1 and that there are n! such T. Harris (1960) offers an alternative development of (4.46) by decomposing the cycle space of T and employing a convenient identity from Katz (1955).

From (4.46) it is straightforward that

$$P(U_n \geq u) = (n)_u/n^u \qquad (4.47)$$

The identity

$$\sum_{u=1}^{n} \frac{(n)_u}{n^u} = \frac{1}{n} \sum_{u=1}^{n} \frac{(n)_u}{n^u} u^2 \quad \text{or} \quad nE\left(\frac{1}{U_n}\right) = E(U_n) \quad (4.48)$$

derived by taking the mean of U_n using (4.46) and equating to (4.25) is thus seen to be a special case of the well-known result that for any variable X on the positive integers,

$$E(X) = \sum_{i=1}^{\infty} P(X \geq i)$$

Continuing in this fashion, from (4.39), we have

$$E(U_n^2) = \sum_{u=1}^{n} (2u-1) \frac{(n)_u}{n^u} = 2n - \sum_{u=1}^{n} \frac{(n)_u}{n^u}$$

or (4.49)

$$E(U_n^2) = 2n - E(U_n)$$

and hence the identity

$$\sum_{u=1}^{n} \frac{(n)_u}{n^u} u^3 = 2n^2 - n \sum_{u=1}^{n} \frac{(n)_u}{n^u} = 2n^2 - \sum_{u=1}^{n} \frac{(n)_u}{n^u} u^2 \quad (4.50)$$

Note that $n^{-1}E(U_n^2) \to 2$.

We next examine the exact distributions of V_n and of $T_{n,r}$ [equivalently $S_{n,r}$ since $P(S_{n,r} = kr) = P(T_{n,r} = k)$]. In each case we do this by conditioning on U_n. The conditional probabilities will involve events such as v cycles resulting from u cyclic states or k cycles of length r resulting from u cyclic states (ignoring the character of cycles not of length r). Such events may be viewed in terms of cycle classes of permutations of u distinct elements. The latter literature is considerable (see Riordan, 1958, Chap. 4).

For example, consider V_n.

$$P(V_n) = v) = \sum_{u=v}^{n} P(V_n = v \mid U_n = u) P(U_n = u)$$

$$= \sum_{u=v}^{n} \alpha_n(u,v) \frac{(n)_u}{n^{u+1}} u \tag{4.51}$$

It is clear that α does not depend on n. It is merely the probability of exactly v cycles resulting from u cyclic elements. Using straightforward generating function arguments (Riordan, 1958, pp. 70-71), we may show that

$$\alpha(u,v) = (-1)^{u+v} s(u,v)/u! \tag{4.52}$$

where s(u,v) are Stirling numbers of the first kind. From the familiar recursive relationship for such Stirling numbers,

$$s(u,v) = s(u-1, v-1) - s(u-1, v)$$

we obtain recursively for $\alpha(u,v)$,

$$\alpha(u,v) = \frac{1}{n} \alpha(u-1, v-1) + \frac{u-1}{u} \alpha(u-1, v) \tag{4.53}$$

In particular,

$\alpha(1,1) = 1$

$\alpha(2,1) = 1/2$ $\quad\alpha(2,2) = 1/2$

$\alpha(3,1) = 1/3$ $\quad\alpha(3,2) = 1/2$ $\quad\alpha(3,3) = 1/6$

and

$\alpha(u,1) = 1/u$ $\quad\alpha(u,u) = 1/u!$

The distribution of V_n is obtained in a more complicated form than (4.51) by Folkert (1955) using the aforementioned Katz identity. Using (4.26) and (4.51) the identity (4.54) ensues.

$$\sum_{u=1}^{n} \frac{1}{u} \frac{(n)_u}{n^u} = \sum_{v=1}^{n} \sum_{u=v}^{n} \alpha(u,v) \frac{(n)_u}{n^{u+1}} uv$$

$$= \sum_{u=1}^{n} \frac{(n)_u}{n^u} \frac{u}{n} \sum_{v=1}^{u} v\alpha(u,v) \tag{4.54}$$

The factorial moments of the conditional distribution of V_n given U_n may be readily obtained. Again a generating function argument (Riordan, 1958, pp. 71-72) shows that $(w \geq 1)$

$$\sum_{v=w}^{u} (v)_w \alpha(u,v) t^{v-w} = \frac{1}{u!} \frac{\partial^w [t(t+1) \ldots (t+u-1)]}{\partial t^w} \tag{4.55}$$

Hence

$$E[(V_n)_w | U_n = u] = \begin{cases} \frac{1}{u!} \left. \frac{\partial^w [t(t+1) \ldots (t+u-1)]}{\partial t^w} \right|_{t=1}, & w \leq u \\ 0, & w > u \end{cases} \tag{4.56}$$

At $w = 1$ we obtain

$$E(V_n | U_n = u) = \left(1 + \frac{1}{2} + \ldots + \frac{1}{u}\right) = \sum_{r=1}^{u} \frac{1}{r} \tag{4.57}$$

Thus the conditional mean of V_n given U_n behaves like $\log U_n$ when U_n is large. In fact,

$$\log(U_n + 1) \leq E(V_n | U_n) \leq \log(U_n + 1) + 1$$

At $w = 2$ we obtain

$$E\left[V_n(V_n - 1) | U_n = u\right] = \sum_{i=1}^{u} \sum_{\substack{j=1 \\ i \neq j}}^{u} \frac{1}{ij} \tag{4.58}$$

whence

$$\operatorname{var}(V_n | U_n = u) = \sum_{\substack{i=1 \\ i \neq j}}^{u} \sum_{j=1}^{u} \frac{i}{ij} + \sum_{i=1}^{u} \frac{1}{i} - \left(\sum_{i=1}^{u} \frac{1}{i} \right)^2$$

$$= \sum_{i=1}^{u} \frac{1}{i} - \sum_{i=1}^{u} \frac{1}{i^2} \tag{4.59}$$

Thus the conditional variance of V_n given U_n also behaves like log U_n when U_n is large. The reader may note the similarities between (4.57) and (4.26) and between (4.59) and (4.40). In fact, using (4.26) and (4.57), we have the identity

$$\sum_{r=1}^{n} \frac{1}{r} \frac{(n)_r}{n^r} = \sum_{u=1}^{n} \frac{(n)_u}{n^{u+1}} u \bullet \sum_{r=1}^{u} \frac{1}{r}$$

which is also seen using (4.47) and interchanging order of summation.

Consider the exact distribution of $T_{n,r}$.

$$P(T_{n,r} = k) = \sum_{n=kr}^{n} P(T_{n,r} = k | U_n = u) \; P(U_n = u)$$

$$= \sum_{n=kr}^{n} \beta_n(u,r,k) \bullet \frac{(n)_u}{n^{u+1}} \tag{4.60}$$

Now β does not depend on n. It is merely the probability of exactly k cycles of length r resulting from u cyclic elements ignoring the cyclic character of the remaining u - kr cyclic elements. It is straightforward to show that

$$\beta(u,r,k) = \frac{1}{k!r^k} \beta(u - kr, r, 0) \tag{4.61}$$

Since $\beta(w,r,0) = 1 - \sum_{k=1}^{[w/r]} \beta(w,r,k)$ ($[x]$ denotes the greatest integer $\le x$) and since $\beta(w,r,0) = 1$ when $w < r$, $\beta(u,r,k)$ can be obtained recursively. Also, $\beta(r,r,1) = \alpha(r,1) = 1/r$ and $\beta(r,1,r) = \alpha(r,r) = 1/r!$.

The factorial moments of the conditional distribution of $T_{n,r}$ given U_n may be readily obtained. Again a generating function argument (Riordan, 1985, pp. 80-81) shows that

$$\sum_{k=0}^{\lceil \frac{u}{r} \rceil} \beta(u,r,k)t^k = \sum_{k=0}^{\lceil \frac{u}{r} \rceil} \frac{1}{k!r^k} (t-1)^k \tag{4.62}$$

Differentiating both sides w times with respect to t yields

$$\sum_{k=w}^{\lceil \frac{u}{r} \rceil} (k)_w \beta(u,r,k)t^{k-w} = \sum_{k=w}^{\lceil \frac{u}{r} \rceil} \frac{(k)_w}{k!r^k} (t-1)^{k-w} \tag{4.63}$$

Evaluating (4.63) at $t = 1$, we have

$$E[(T_{n,r})_w | U_n = u] = \begin{cases} r^{-w}, & w \le \lceil \frac{u}{r} \rceil \\ 0, & w > \lceil \frac{u}{r} \rceil \end{cases} \tag{4.64}$$

Hence

$$E(T_{n,r} | U_n = u) = \frac{1}{r}, \quad r \le u \tag{4.65}$$

$$\mathrm{var}(T_{n,r} | U_n = u) = \frac{1}{r}, \quad r \le u/2 \tag{4.66}$$

Again the reader may note the similarities between (4.65) and (4.23) and between (4.66) and (4.36). In fact, from (4.65)

$$E(T_{n,r}) = \sum_{u=r}^{n} \frac{1}{r} \frac{(n)_u}{n^{u+1}} u = \frac{1}{r} \frac{(n)_r}{n^{r+1}}$$

[using (4.47)], in agreement with (4.23). Also, we may sum both sides of (4.65) up to u to obtain (4.57).

In concluding this section we examine the expected length of a cycle denoted by ECL. We first compute the likelihood of any particular cycle space configuration under a random T. If we let m be the number of cycles of length l, $l = 1, \ldots, n$, and let $m_o = n - \Sigma m$ = the number of transient states then (with $m_o \geq 0$)

P(m_l cycles of length l and m_o transient states)

= P(m_l cycles of length $l | U_n = n-m_o$) $P(U_n = n-m_o)$

$$= \frac{1}{\prod_{l=1}^{n-m_o} m_l! \prod_{l=1}^{n-m_o} l^{m_l}} \bullet \frac{(n)_{n-m_o}(n-m_o)}{n^{n-m_o+1}}$$

$$= \frac{n!}{\prod_{l=1}^{n-m_o} m_l! \prod_{l=1}^{n-m_o} l^{m_l}} \cdot \frac{n-m_o}{n^{n-m_o+1}} \equiv P(m_o, m_1, \ldots, m_n)$$

Sherlock (1979) provides discussion of the conditional distribution of cycle configuration given $U_n = u$ and verifies that

$$\sum_{(m_1,\ldots,m_u)\varepsilon S_u} \left(\prod_{l=1}^{u} m_l! \prod_{l=1}^{u} l^{m_l} \right)^{-1} = 1$$

where, as before, S_u denotes the set of all partitions of u. In fact,

$$\sum_{S_u} m_r \left(\prod_{l=1}^{u} m_l! \prod_{l=1}^{u} l^{m_l} \right)^{-1} = 1/r \tag{4.67}$$

a result speculated by Sherlock (1979, p. 697) which is apparent from (4.65) since the left-hand side of (4.67) is precisely $E(T_{n,r} | U_n = u)$.

Given any vector $m_1, \ldots, m_n$ such that $m_l \geq 0$ and $\Sigma m_l \leq n$, the average cycle length for the cycle space configuration it defines is $(\Sigma m_l)^{-1} \Sigma m_l l$. Hence

$$ECL = \Sigma[(\Sigma m_l)^{-1} \Sigma m_l l \cdot P(m_o, m_1, \ldots, m_n)] \quad (4.68)$$

where the outer sum is over the set $\{(m_1, m_2, \ldots, m_n): \Sigma m_l l \leq n, m_l \geq 0\}$.

Continuing, we note that $\Sigma m_l l$ is a value of U_n and Σm_l is a value of V_n and thus

$$ECL = E\left(\frac{U_n}{V_n}\right)$$

Using the joint distribution of U_n, V_n contained in (4.51), we have

$$\begin{aligned} ECL &= \sum_{v=1}^{n} \sum_{u=v}^{n} \frac{u}{v} \alpha(u,v) \frac{(n)_u}{n^{u+1}} u \\ &= \sum_{u=1}^{n} \sum_{v=1}^{n} \frac{u^2}{v} \alpha(u,v) \frac{(n)_u}{n^{u+1}} \end{aligned} \quad (4.69)$$

From the Cauchy-Schwarz inequality we have

$$ECL = E\left(\frac{U_n}{V_n}\right) \geq [E(\sqrt{U_n})]^2/E(V_n) \quad (4.70)$$

which may be used to show that ECL $\to \infty$ as $n \to \infty$ [see (4.75)].

It is noteworthy that in discussing ECL we have, for a particular net, defined "average cycle length" assuming cycles to be equally likely, for example, if a net has three cycles of lengths 10, 5 and 3, we obtain an "average cycle length" = $10 \cdot \frac{1}{3} + 5 \cdot \frac{1}{3} + 3 \cdot \frac{1}{3} = 6$.

Average cycle length for a particular net may also be defined assuming an equally likely selection of a cyclic state. For the example above we would then obtain an average cycle length $10 \cdot \frac{10}{18} + 5 \cdot \frac{5}{18} + 3 \cdot \frac{3}{18} = 7.4$. Cull (1978) studies ECL under this latter definition.

4.3.3 Asymptotic Results

Using ideas from Harris (1960), we obtain the asymptotic probability density of U_n. Letting $W_n = U_n/\sqrt{n}$ and using (4.46), we may show after some manipulation that W_n converges in distribution to a random variable W having a Rayleigh distribution, that is, the density of W is

$$f_w(w) = we^{-w^2/2}, \quad w > 0 \tag{4.71}$$

This conclusion also establishes that $U \overset{p}{\to} \infty$ ("p" indicates convergence in probability). It is easy to show that

$$E(W^r) = 2^{r/2}\Gamma\left(\frac{r+2}{2}\right), \quad r > -2$$

Thus for $k > -2$,

$$E(n^{-k/2}U_n^k) = n^{-k/2} \sum_{u=1}^{n} \frac{u^{k+1}(n)_u}{n^{u+1}} \to 2^{k/2}\,\Gamma\frac{k+2}{2}$$

In particular from (4.48) we have

$$E\frac{U_n}{\sqrt{n}} = n^{-1/2}\sum_{u=1}^{n}\frac{(n)_u u^2}{n^{u+1}} = n^{-1/2}\sum_{u=1}^{n}\frac{(n)_u}{n^u} \tag{4.72}$$

$$= \frac{E\sqrt{n}}{U_n} \to \sqrt{\pi/2}$$

so that $E(U_n) \approx \sqrt{n}\sqrt{\pi/2}$ [i.e., $E(U_n) = O(\sqrt{n})$], verifying the limit (4.24). Furthermore, in agreement with our remark after (4.50), we have

$$n^{-1}E(U_n^2) = n^{-1} \sum_{u=1}^{n} \frac{(n)_u}{n^{u+1}} u^3 \to E(W^2) = 2$$

Expression (4.72) also implies that the expected number of transient states approaches ∞ as $n \to \infty$, that is,

$$E(n - U_n) = \sqrt{n}\, E(\sqrt{n} - \frac{U_n}{\sqrt{n}}) \to \infty$$

Additionally, $\mathrm{var}(U_n/\sqrt{n}) \to 2 - \pi/2$, confirming that $\mathrm{var}(U_n) \to \infty$, as noted after (4.40).

As for V_n, expression (4.57) provides a verification of the limit in (4.26). Since $U_n \overset{p}{\to} \infty$, $\log U_n \overset{p}{\to} \infty$ and thus $E(\log U_n) \overset{p}{\to} \infty$. Therefore,

$$E(V_n) = EE(V_n|U_n) \geq E[\log(U_n + 1)] \to \infty$$

More precisely,

$$\begin{aligned} E(V_n) &\approx E \log (U_n) \\ &\approx \log E(U_n) - \mathrm{var}(U_n)/2[E(U_n)]^2 \\ &\approx \frac{\log n}{2} + \log \sqrt{\pi/2} - \frac{2 - \pi/2}{\pi} \end{aligned}$$

so that

$$(\log n)^{-1}E(V_n) \to 1/2$$

or

$$E(V_n) \approx \frac{\log n}{2} \quad [E(V_n) = O(\log n)] \tag{4.73}$$

Similarly, expression (4.59) shows that $\mathrm{var}(V_n) \to \infty$:

$$\mathrm{var}(V_n) \geq E[\mathrm{var}(V_n)|U_n)] \geq E[\log(U_n + 1)] - 2 \to \infty$$

As above, $\mathrm{var}(V_n)$ may be approximated.

In support of these findings, we see that from (4.40)

$$E(V_n^2) = \sum_{r=1}^{n} \frac{1}{r} \frac{(n)_r}{n^r} + \sum_{\substack{r,s>1 \\ r+s>n}} \frac{1}{r}\frac{1}{s} \frac{(n)_{r+s}}{n^{r+s}}$$

$$\leq E(V_n) + E(U_n)$$

$$\leq 2E(U_n) \text{ since } V_n \leq U_n$$

Hence, again using (4.72), $n^{-1}E(V_n^2) \to 0$, implying that $n^{-1/2}E(V_n) \to 0$ and that $\mathrm{var}(\sqrt{n}\ V) \to 0$.

We now demonstrate that as $n \to \infty$, ECL $\to \infty$. The inequality (4.70) may be written as

$$\mathrm{ECL} \geq [E(n^{-1/4}U_n^{-1/2})]^2\ [E(n^{-1/2}V_n)]^{-1} \tag{4.74}$$

Using (4.72) and (4.73), the right-hand side of (4.74) clearly tends to ∞ as n does. More precisely,

$$\mathrm{ECL} = E\left[\frac{U_n}{V_n}\right] \approx \frac{E(U_n)}{E(V_n)} - \frac{\mathrm{cov}(U_n,V_n)}{[E(V_n)]^2} + \frac{\mathrm{var}(V_n)E(U_n)}{2[E(V_n)]^3}$$

$$\approx \frac{\sqrt{n}\ \sqrt{2\pi}}{(\log\ n)/2} + O\left[\frac{\sqrt{n}}{(\log\ n)^2}\right]$$

The order $\sqrt{n}/(\log\ n)^2$ is seen since $\mathrm{cov}(U_n,V_n)$ is $O(\sqrt{n})$ and $\mathrm{var}(V_n)$ is $O(\log\ n)$. Thus

$$\frac{\log\ n}{\sqrt{n}}\ \mathrm{ECL} \to 2\sqrt{2\pi}$$

or

$$\mathrm{ECL} \approx \frac{\sqrt{n}}{\log\ n}\ 2\sqrt{2\pi} \tag{4.75}$$

that is,

$$\mathrm{ECL} = O\left(\frac{\sqrt{n}}{\log\ n}\right)$$

[although the $O(\sqrt{n}/(\log\ n)^2)$ term is very slowly asymptotically negligible)].

We next argue that the asymptotic distribution of $T_{n,r}$ (i.e., $S_{n,r}/r$) is Poisson with mean $1/r$. The limits in (4.23) and (4.35) encourage this conclusion. It suffices to show that

$$\lim_{n\to\infty} [k!r^k P(T_{n,r} = k) - P(T_{n,r} = 0)] = 0, \quad k = 1, 2, \ldots \tag{4.76}$$

From (4.60) and (4.61) we may write

$$P(T_{n,r} = k) = \frac{1}{k!r^k} \sum_{u=0}^{n-kr} \beta(u,r,0) \frac{(n)_{u+kr}(u + kr)}{n^{u+kr+1}}$$

Hence the left-hand side of (4.76) becomes

$$\lim_{n\to\infty} \left[\sum_{u=1}^{n} \beta(u,r,0) \frac{(n)_u u}{n^{u+1}} - \sum_{u=0}^{n-kr} \beta(u,r,0) \frac{(n)_{u+kr}(u+kr)}{n^{u+kr+1}} \right]$$

$$= \lim_{n\to\infty} \left[\sum_{u=1}^{n-kr} \beta(u,r,0)u \left(\frac{(n)_n}{n^{u+1}} - \frac{(n)_{u+kr}}{n^{u+kr+1}} \right) \right.$$

$$\left. + \sum_{u=n-kr+1}^{n} \beta(u,r,0) \frac{(n)_u u}{n^{u+1}} - kr \sum_{u=0}^{n-kr} \frac{(n)_{u+kr}}{n^{u+kr+1}} - \frac{(n)_{kr}(kr)}{n^{kr+1}} \right]$$

It is apparent that the limits of the second, third, and fourth terms within the brackets are 0. Since $\beta(u,r,0) \le 1$ and since

$$\lim_{n\to\infty} \sum_{u=a}^{n} \frac{(n)_u u}{n^{u+1}} = \lim_{n\to\infty} \sum_{u=1}^{n-a} \frac{(n)_u u}{n^{u+1}} = 1$$

for any fixed positive integer a, the first term also tends to 0 and we are done.

A simpler derivation arises from the well-known fact that if

$$X \sim P_o(\lambda) \text{ then } E(X)_w = \lambda^w \tag{4.77}$$

(see e.g., Johnson and Kotz, 1969, p. 90). Hence we need only show that $\lim_{n\to\infty} E[(T_{n,r})_w] = r^{-w}$, $w = 1, 2, \ldots$. Using (4.64) and the fact that $U_n \to \infty$, we have

$$\lim_{n\to\infty} E[(T_{n,r})_w] = \lim_{n\to\infty} E[E(T_{n,r})_w | U_n)]$$

$$= \lim_{n\to\infty} r^{-w} P\left(\left[\frac{U_n}{\lfloor rn \rfloor}\right] \geq w\right)$$

$$= r^{-w}$$

Summarizing then, we have $T_{n,r}$ converging in distribution to Poisson$(1/r)$. (The limiting Poisson distribution when $r = 1$ was noted earlier.)

From (4.77) we may also conclude that if $X \sim P_o(\lambda)$,

$$E(X^k) = \sum_{j=1}^{k} S(k,j)\lambda^j$$

where the $S(k,j)$ are Stirling numbers of the second kind. Hence

$$E(T_{n,r})^k \to \sum_{j=1}^{k} S(k,j)r^{-j} \tag{4.78}$$

We calculate the left-hand side of (4.78) assuming that $n > k$:

$$E(T_{n,r})^k = r^{-k} E\left(\sum_{i=1}^{n} D_{ri}^n\right)^k$$

$$= r^{-k} \sum_{R} - \frac{k!}{\prod k_i!} E\prod(D_{ri}^n)^k$$

where $R = \{(k_1, \ldots, k_n): k > 0, k_i = k\}$. But if exactly m of the $k_i \neq 0$, $E\prod(D_{ri})^{k_i} = P_{n,m,r}$, given by (4.42). Continuing then, we have

$$E(T_{n,r})^k = r^{-k} \sum_{m=1}^{k} P_{n,m,r} \sum_{R_m} \frac{k!}{\Pi k_i!}$$

where R_m denotes the subset of R on which exactly m of the k_i's are greater than 0. But the sum over R_m is merely the number of ways of placing k objects into n cells such that exactly m are nonempty. The number is $(n)_m S(k,m)$ (Riordan, 1958, p. 92), whence

$$E(T_{n,r})^k = r^{-k} \sum_{m=1}^{k} S(k,m)(n)_m P_{n,m,r}$$

Using (4.43), we have

$$\lim_{n\to\infty} (E(T_{n,r})^k = r^{-k} \sum_{m=1}^{k} S(k,m) \sum_{S_{m,r}} n(m_1,\ldots,m_j) \quad (4.79)$$

$$[(r-1)!]^j \; [\prod_{i=1}^{j} (r-m)!]^{-1}$$

Denoting the sum over $S_{m,r}$ by $\Delta_{r,m}$ and equating right-hand sides in (4.78) and (4.79), we find the identity

$$\sum_{m=1}^{k} S(k,m)r^{k-m} = \sum_{m=1}^{k} S(k,m)\Delta_{r,m} \quad (4.80)$$

Note that $\Delta_{1,m} = 1$ reduces (4.80) to a triviality.

We briefly summarize the most important asymptotic conclusions of this section in terms of completely random nets. Restoring $n = 2^N$, we have demonstrated that

1. The expected number of cyclic states is of order $2^{N/2}$.

2. The expected number of transient states is of order 2^N.
3. The expected number of cycles is of order N.
4. The likelihood that any particular state is cyclic of order $2^{-N/2}$.
5. The expected number of cycles of length r converges to 1/r.
6. The expected number of states on cycles of length r converges to 1.
7. The expected cycle length is of order $(N^{-1})2^{N/2}$.

4.4 Forcibility and Internal Homogeneity

The material presented in this section and in Secs. 4.5 and 4.6 is drawn from Walker and Gelfand (1977) and Gelfand and Walker (1979). Section 4.3 suggests and empirical evidence in Chapter 5 supports the fact that random nets are behaviorally unattractive as the number of elements N grows large. Hence we must exert control over the cycle space to accomplish fewer cycles and cycles of reasonable length if we are to exhibit switching nets as behaviorally plausible models. In the spirit of Kauffman's remarks (Sec. 3.5.1), we consider local control, that is, control over the individual element responses to input information. This means imposing restrictions on the selection of Boolean transformations (mappings) guiding the individual elements. In this and the next sections we define and characterize several notions of constraint together with their interrelationships. Detailed examination of the behavioral ramifications of these control measures for the entire net requires extensive simulation work and is discussed in Chapter 5.

How does one exert local control? Walker and Ashby (1966) found that increasing the sameness of the entries (i.e., the internal homogeneity) of the elements' function table tends to decrease the length of cycles. Kauffman (1969, 1970) found the number of inputs to be importantly related to both cycle length and to their stability. In the latter paper he describes a powerful explanatory property called forcibility which is also implicated in cyclic behavior. Babcock (1976) has shown that the two measures, internal homogeneity and forcibility, are related.

To formalize these notions, consider a mapping on k inputs. By definition, a mapping is forcible on a given input when a given state of the input "forces" the output of the mapping to a single value regardless of the values of the other inputs. This given state is called the forcing state. If an input is forcing on both states, the mapping is either constant (trivial) or has half 1's and half 0's. In the former case all inputs are forcing on both states, while in the latter case the mapping must be forcing only on the one input. Since forcibility with only one input is trivial, we restrict attention to the case where the number of inputs $k \geq 2$. The forced value of an element is that value to which it is forcible.

If an element is forcible on more than one input line, its forced value is identical for all the inputs on which it is forcible. This is apparent since forcibility on a particular input implies that the mapping assumes a forced value on at least 2^{k-1} of the 2^k input vectors. It is easy to verify that of the 16 mappings on two input coordinates, 10 are forcible on both coordinates (including the two trivial constant mappings), 4 are forcible on one, and 2 are forcible on neither. Notice that forcibility as it is formulated is a "local" property; we examine mappings at individual elements in a system. In Chapter 5 we consider ramifications of forcibility for the entire net.

Internal homogeneity, denoted henceforth by I, is defined as the larger of the number of 0 entries and of 1 entries in the table of values of a mapping, that is, I = max(no. 0's, no. 1's), and hence $2^{k-1} \le I \le 2^k$. We denote by $N_k(i)$ the number of mappings with I = i.

We now enumerate the forcible mappings. The major result is Theorem 2. To develop this result, we begin with several elementary lemmas.

Lemma 1: $N_k(i) = 2\binom{2^k}{i}$ for $2^{k-1} < i \le 2^k$ and if $I = 2^{k-1}$,

$$N_k(2^{k-1}) = \binom{2^k}{2^{k-1}}$$

Proof: Obvious.

Corollary 1: The mean internal homogeneity for mappings on k inputs is

$$2^{k-1} + 2^{k+1-2^k}\binom{2^k - 1}{2^{k-1} - 1} \tag{4.81}$$

Proof: We seek

$$2^{-2^k}\sum_{i=2^{k-1}}^{2^k} i\,N(i) = 2^{-2^k}\left[\sum_{i=2^{k-1}+1}^{2^k} i \bullet 2\binom{2^k}{i} + 2^{k-1}\binom{2^k}{2^{k-1}}\right]$$

$$= 2^{-2^k}\left[2^k \bullet \sum_{i=2^{k-1}+1}^{2^k} 2\binom{2^k-1}{i-1} + 2^k\binom{2^k-1}{2^{k-1}-1}\right]$$

$$= 2^{-2^k}\left[2^k \bullet \sum_{j=2^{k-1}}^{2^k-1} 2\binom{2^k-1}{j} + 2^k\binom{2^k-1}{2^{k-1}-1}\right]$$

$$= 2^{-2^k}\left[2^k \bullet \sum_{j=0}^{2^k-1} \binom{2^k-1}{j} \quad 2^{k+1}\binom{2^k-1}{2^{k-1}-1}\right]$$

$$= 2^{-2^k}\left[2^k \bullet 2^{2^k-1} + 2^{k+1}\binom{2^k-1}{2^{k-1}-1}\right]$$

$$= 2^{k-1} + 2^{k+1-2^k}\binom{2^k-1}{2^{k-1}-1}$$

We wish to enumerate $\Gamma(k,i,j)$ which is defined to be the number of mappings on k inputs with internal homogeneity i, which are forcing on *exactly* j of the k inputs. We will develop a recursive procedure which calculates $\Gamma(k,i,j)$ for $j > 0$. We can then obtain $\Gamma(k,i,0)$ by

$$\Gamma(k,i,0) = N_k(i) - \sum_{j=1}^{k} \Gamma(k,i,j) \tag{4.82}$$

From $\Gamma(k,i,j)$ we can calculate the number of mappings on k inputs with internal homogeneity i which are forcing on at least (at most, etc.) j inputs by simple summation. Moreover, we can calculate the number of mappings on k inputs forcing on exactly j inputs, $\Gamma(k,j)$ again by summation, that is,

$$\Gamma(k,j) = \sum_{i=2^{k-1}}^{2^k} \Gamma(k,i,j) \qquad (4.83)$$

If we seek the density of any of these collections of mappings, we standardize by either $N_k(i)$ or 2^{2^k}, depending on the base of reference.

First we pose the following questions. Given k and j, how large must i be to allow the possibility of j forcing inputs? Reciprocally, given k and i, how many forcing inputs are possible; that is, how large may j be? The answer is expressed in the following lemma.

Lemma 2:

(i) Given k and j, we cannot have j forcing inputs unless $i \geq 2^k - 2^{k-j}$.

(ii) Given k and i, a mapping may have at most $[k - \log_2(2^k - i)]$ forcing inputs (again [x] denotes the greatest integer $\leq x$).

Proof: Result (ii) follows from result (i) by solving the inequality for j so that we need only establish (i). To see (i) we note that in order that a mapping be forcing on one input, we need at least half of the entries to assume the forced value. From the lexicographic order, if the mapping is to be forcing on a second input, we need at least half of the remaining entries to assume the forced value. Building inductively, we need $i \geq 2^{k-1} + 2^{k-2} + \ldots + 2^{k-j}$ in order that we have a sufficient number of entries at the forced value to allow j forcing inputs. Simplifying this inequality produces the result in (i).

For convenience, we set $\Gamma(k,i,j) = 0$ if $i < 2^k - 2^{k-j}$. We can now calculate $\Gamma(k,i,j)$. In the following lemma we take care of the two extreme cases.

Lemma 3:

$$\text{(i)} \quad \Gamma(k,2^k,j) = \begin{cases} 2, & j = k \\ 0, & j \neq k \end{cases}$$

$$\text{(ii)} \quad \Gamma(k,2^{k-1},j) = \begin{cases} 2k, & j = 1 \\ \binom{2^k}{2^{k-1}} - 2k, & j = 0 \\ 0, & \text{otherwise} \end{cases}$$

Proof: (i) Obvious. (ii) Again considering the lexicographic order, there will be exactly two mappings forcing on the first input; that is, the upper half 1's and the lower half 0's, or vice versa. But the selection of the first input is but one of k choices, so in total $\Gamma(k,2^{k-1},1) = \binom{k}{1}2 = 2k$. It is impossible by Lemma 2 to have more than one forcing input (this is clear directly) and thus

$$\Gamma(k,2^{k-1},0) = N_k(2^{k-1}) - 2k = \binom{2^k}{2^{k-1}} - 2k$$

by Lemma 1, we now proceed to the major result.

Theorem 2: If $2^{k-1} < i < 2^k$ and $j > 0$,

$$\Gamma(k,i,j) = \binom{k}{j} 2^{j+1} \gamma(k,i,j)$$

where

$$\gamma(k,i,j) = \begin{cases} \binom{2^{k-1}}{2 - i} & \text{if } i < 2^k - 2^{k-j} + 2^{k-j-1} \\ \dfrac{\Gamma(k-j, i-2^k+2^{k-j}, 0)}{2} & \text{if } i > 2^k - 2^{k-j} + 2^{k-j-1} \\ \binom{2^{k-j}}{2^{k-j-1}} - 2(k-j) & \text{if } i = 2^k - 2^{k-j} + 2^{k-j-1} \end{cases}$$

[Note: Conventionally we take $\Gamma(1,1,0) = 0$, $\Gamma(1,1,1) = 2$ for consistency of notation since we have restricted $k \geq 2$. In fact, these conventions coincide with our definitions in this trivial case.]

Proof: In our proof we consider the first j inputs. Hence the $\binom{k}{j}$ term is needed to adjust for the arbitrary selection of j inputs. The 2^{j+1} term arises from there being two choices for the forced value together with 2^j possibilities for the j forcing inputs in selecting which value for each of the inputs forces to the forced value.

The γ term provides the most crucial counting. As in the argument for Lemma 1, we will need $2^k - 2^{k-j}$ entries at the forced value in order that the first j inputs will be forcing. The remaining entries at the forced value must be arranged to force *none* of the remaining inputs. Hence the problem is collapsed to examining mappings on a reduced number of inputs k - j with $i - 2^k + 2^{k-j}$ entries at the forced value and 2^{k-1} at the nonforced value. If

$$i - 2^k + 2^{k-j} < \frac{1}{2}(2^{k-j}) = 2^{k-j-1}$$

any arrangement of these entries will fail to force another input, that is, $\Gamma(k,i,j) = \binom{2^{k-j}}{2^k - i}$. If

$$i - 2^k + 2^{k-j} > \frac{1}{2}(2^{k-j}) = 2^{k-j-1}$$

to ensure that these entries will not force another input, we need

$$\Gamma(k,i,j) = \frac{\Gamma(k-j, i-2^k+2^{k-j}, 0)}{2}$$

(The reason for the divisor of 2 is that we do not at this point have two choices for the forced value which we had in the original counting.) Finally, if

$$i-2^k + 2^{k-j} = \frac{1}{2}(2^{k-j}) = 2^{k-j-1}$$

we are precisely in the situation of Lemma 3 and need

$$\Gamma(k-j, 2^{k-j-1}, 0) = \binom{2^{k-j}}{2^{k-j-1}} - 2(k-j)$$

Rearranging the inequalities on i, k, and j produces the exact form of the theorem.

This construction enables recursive calculation of $\Gamma(k,i,j)$. The tabulations given in Table 4.1 may be verified both directly and from Theorem 2. The row totals provide $\Gamma(k,j)$, while the column totals provide $N_k(i)$.

Corollary 2: (i) $\Gamma(k, 2^k-1, k) = 2^{k+1} = N_k(2^k-1)$

(ii) $\Gamma(k, 2^k-2, k-1) = k \cdot 2^k = \Gamma(k, k-1)$

Proof: (i) and the left equality in (ii) are obvious by calculation. The right equality in (ii) follows since Lemma 2 requires that $i \geq 2^k - 2$ in order to allow $k - 1$ forcing inputs and (i) shows $\Gamma(k, 2^k - 1, k - 1) = 0$.

(i) of Corollary 2 implies that any mapping on k inputs with internal homogeneity equal to $2^k - 1$ will be forcing on all inputs; (ii) implies that the only mappings on k inputs which will be forcing on $k - 1$ of them have internal homogeneity equal to $2^k - 2$. A result of the latter type has been discussed by Newman and Rice (1971).

For analytical purposes the results of Theorem 2 are a bit awkward to use. Since $\Gamma(k,i,0)$ is calculated by subtraction and yet is also needed for subsequent $\Gamma(k,i,j)$, it becomes very difficult to employ this theorem to bound $\Gamma(k,i,j)$ and hence $\Gamma(k,j)$.

We seek a simple bound on the number of mappings forcing on at least j inputs. In our notation we are bounding

Table 4.1 Γ(k,i,j) for k = 2, 3, 4

k = 2

j \ i	2	3	4	
0	2	0	0	2
1	4	0	0	4
2	0	8	2	10
	6	8	2	16

k = 3

j \ i	4	5	6	7	8	
0	64	64	8	0	0	136
1	6	48	24	0	0	78
2	0	0	24	0	0	24
3	0	0	0	16	2	18
	70	112	56	16	2	256

k=4

j \ i	8	9	10	11	12	13	14	15	16	
0	12,862	22,752	15,568	7,840	2,568	640	16	0	0	62,246
1	8	128	448	896	1,024	288	64	0	0	2,856
2	0	0	0	0	48	192	96	0	0	336
3	0	0	0	0	0	0	64	0	0	64
4	0	0	0	0	0	0	0	32	2	34
	12,870	22,880	16,016	8,736	3,640	1,120	240	32	2	65,536

$$\sum_{j'=j}^{k} \Gamma(k,j') \tag{4.84}$$

A convenient result is given in Theorem 3.

Theorem 3: $\displaystyle\sum_{j'=j}^{k} \Gamma(k,j') \leq 2^{j+1} \binom{k}{j} 2^{2^{k-j}}$

Proof: The 2^{j+1} and $\binom{k}{j}$ terms arise for the same reasons as in the proof of Theorem 2. We may therefore focus on the first j inputs and assume that 1 is the forcing input value and that 1 is the forced value as well.

If the first $2^{k-1} + 2^{k-2} + \ldots + 2^{k-j} = 2^k - 2^{k-j}$ of the 2^k entries in the mapping are at the forced value 1, then regardless of the remaining 2^{k-j} entries, we will have at least j forcing inputs. (Recall the proof of Lemma 2.) These remaining entries may be selected in $2^{2^{k-j}}$ ways, completing the result.

Much double counting is involved in obtaining this upper bound. Each mapping forcing on, say, $j' > j$ inputs will be counted once for each of the $\binom{j'}{j}$ subsets of j inputs.

When $j = 1$ we have a bound on the total number of mappings forcible on one or more input line. Kauffman (1970, p. 105) offers an incorrect bound in his Theorem 2. The sums should be replaced by products. His subsequent Theorems 3, 4, and 5 suffer from the same error. In any event the bound from Theorem 3, which is

$$4k \cdot 2^{2^{k-1}} \tag{4.85}$$

is easily seen to be tighter than his corrected bound.

This bound enables us to show (as suggested by Table 4.1) that the density of mappings not forcing on any inputs tends to 1 as k increases; that is, the density of forcible maps goes to zero as k grows large.

$$\frac{\sum_{j=1}^{k} \Gamma(k,j')}{2^{2^k}} \leq \frac{4k \cdot 2^{2^{k-1}}}{2^{2^k}} = \frac{4k}{2^{2^{k-1}}} \rightarrow 0$$

as $k \rightarrow \infty$, and we are done.

We restate this result as a corollary.

Corollary 3: The density of forcible maps (i.e., maps forcing on at least one input line) tends to zero as the number of inputs increases.

In concluding this section we obtain a companion result to Corollary 3 by bounding the mean number of forcing inputs for mappings on k inputs.

Corollary 4: The mean number of forcing inputs for mappings on k inputs is at most

$$2^{-2^{k-1}} \cdot 4k^2$$

Proof: We seek

$$2^{-2^k} \sum_{j=0}^{k} j\Gamma(k,j)$$

$$\leq 2^{-2^k} k \sum_{j=1}^{k} \Gamma(k,j)$$

$$\leq 2^{-2^k} k \cdot 4k \cdot 2^{2^{k-1}} = 2^{-2^{k-1}} \quad 4k^2$$

If maps are classed by k, sharp differences are found to exist, particularly for forcibility. From Table 4.1 it can be seen that the density of forcible maps drops sharply as k is increased, from 0.875 at k = 2 (and from 1.0 at k = 1) to 0.05 at k = 4. From

Corollary 3, the limiting density is 0. The mean fraction of forcing inputs per map declines at approximately the same rate: from all inputs forcing at k = 1 to about 1.5% of inputs forcing at k = 4. In fact, from Corollary 4 we see that the mean fraction of forcing inputs for mappings on k inputs is at most $k^{-1} \cdot 2^{-2} \cdot 4k^2$, which approaches zero very quickly as k increases. The drop is sharp enough to justify a simple summary: classified by number of inputs, only Boolean maps with fewer than four inputs are forcible. That is, if mappings are selected equiprobably, nets using elements with four or more inputs will have very low densities of forcing elements.

An important final point is that Table 4.1, together with several of the theorems and lemmas, demonstrates a direct relationship between internal homogeneity and number of forcing inputs.

4.5 Extended Threshold

We define the notation of a mapping on k inputs which has (extended) threshold $l, 1 \leq l \leq k$. Speaking casually, we may say that a mapping on k inputs has threshold l if whenever l or more inputs take on a specified value, the mapping takes on a specified value. We call the resultant mapping or output value the *threshold state* associated with that input value. The specified input value may be 0 or 1 and may be coupled with a threshold state of 0 or 1. In this definition l is the minimum number of inputs for which the statement is true since if the statement holds at l it would obviously hold at $l + 1$, $l + 2$, ..., k.

The difficulty associated with such an informal definition may be revealed by attempting to answer the following illustrative question. Can the system ever be "on" if fewer than l inputs are "on"? If the answer is no, we shall refer to l as an absolute threshold although it is not clear whether l or $k - l$ should be called the threshold since $k - l$ inputs "off" imply that the system is "off." If the answer is yes, we shall refer to l only as a threshold (for the number of "on" inputs). Every mapping must have a threshold (at the largest it would be k), but only a subset of mappings have an absolute threshold. Because our extended conception of a threshold allows either "off" or "on" inputs to turn an element again either "off" or "on," how do we assign a threshold value, l, to a mapping? We need to consider for a mapping m a threshold for the number of "off" or 0 inputs which we denote by $l_0(m)$ and similarly a threshold for the number of "on" or 1 inputs which we denote by $l_1(m)$. We then define $l(m) = \min[l_0(m), l_1(m)]$. In light of our extended definition, the minimum of these two numbers is clearly the more significant value. In considering absolute thresholds we quickly discover that if l_0 is an absolute threshold, so is l_1 and that $l_0 + l_1 = k + 1$. We will prove shortly that for any mapping m, $l_0(m) + l_1(m) \geq k + 1$. Table 4.2 attempts to clarify the definitions and notations. For mapping m_1 we have $l_0(m_1) = 4$, $l_1(m_1) = 3$, and $l(m_1) = 3$. For mapping m_2 we have $l_0(m_2) = 2$, $l_1(m_2) = 4$, and $l(m_2) = 2$. For mapping m_3 we have $l_0(m_3) = 2$, $l_1(m_3) = 3$ with both l_0 and l_1 absolute thresholds and $l(m_3) = 2$. Finally, for mapping m_4 we have a situation where the notation of a threshold has little significance, that is, $l_0(m_4) = l_1(m_4) = l(m_4) = 4$. For the two trivial or constant mappings we define $l = 0$ and do not consider them further in this discussion.

A brief examination of Table 4.2 reveals that the lexicographic ordering for a mapping is not at all convenient for establishing thresholds. A better

Table 4.2 Illustrative Mappings on Four Inputs

x_4	x_3	x_2	x_1	m_1	m_2	m_3	m_4
1	1	1	1	1	1	1	0
0	1	1	1	1	0	1	1
1	0	1	1	1	1	1	0
0	0	1	1	0	0	0	1
1	1	0	1	1	1	1	1
0	1	0	1	0	0	0	0
1	0	0	1	0	0	0	1
0	0	0	1	1	0	0	0
1	1	1	0	1	1	1	1
0	1	1	0	0	0	0	0
1	0	1	0	0	0	0	1
0	0	1	0	1	0	0	0
1	1	0	0	1	0	0	0
0	1	0	0	0	0	0	0
1	0	0	0	0	0	0	1
0	0	0	0	1	0	0	0

arrangement would be to order the input rows monotonically by the number of 0's (hence by the number of 1's). To obtain l_0 and l_1 from this "monotonic" ordering is simple. Suppose, for example, that the input rows are arranged by *increasing* number of 0's. If we scan up the mapping value column in this table for the first change of value (0 to 1 or 1 to 0) and it occurs for a row having j input 0's, then $l_0 = j + 1$. If we scan down the mapping value column for the changes and it occurs for a row having j' input 0's (hence k - j' input 1's), then $l_1 = k - j' + 1$.

We may immediately notice that since we are considering only nontrivial maps we must have $j' < j + 1$

(i.e., $k - l_1 + 1 < l_0$ or $l_0 + l_1 > k = 1$). As Lemma 4 we state and formally prove a slightly broader result.

Lemma 4: For any nontrivial mapping m, $l_0(m) + l_1(m) \geq k + 1$ with equality iff l_0 (and hence l_1) is an absolute threshold.

Proof: If for m, $l_0(m) = j$, then all the input rows having j or more 0's will be forced to a common mapping value. The remaining rows have at most $j - 1$ 0's hence at least $k - j + 1$ 1's. Among rows have $j - 1$ 0's there must be at least one row with mapping value different from the common mapping value for all input rows with j or more 0's. Hence $l_1(m) \geq k - j + 1$ and thus $l_0(m) + l_1(m) \geq k + 1$. Finally, $l_1(m) = k - j + 1$ (i.e., $l_0(m) + l_1(m) = k + 1$) iff all rows with $j - 1$ or fewer 0's have a common mapping value. This mapping clearly has $l_0(m) = j$ and $l_1(m) = k - j - 1$ as absolute thresholds.

We note that if $l(m) = 1$, then l is immediately an absolute threshold. Mappings have threshold $l_0 = 1$ and corresponding threshold state 1 have been defined as noncontractible by Rosen (1958). We thus speak more generally of mappings with $l = 1$ as being extended noncontractible.

We now attempt to enumerate the number of mappings having a particular threshold l. If l is an absolute threshold, the counting is simple. Given l_0 and hence $l_1 = k + 1 - l_0$, there will be two mappings having this particular l_0 and l_1--the mapping having threshold state 0 on the 0 inputs, threshold state 1 on the 1 inputs, and vice versa. Since l_0 runs from 1 to k, the total number of mappings having an absolute threshold is 2k. The number having absolute threshold l is 4 if $l_0 \neq l_1$ and 2 if $l_0 = l_1$. Note that $l_0 = l_1$ implies that k is odd and $l = (k + 1)/2$. Only for this l may the number of maps having absolute threshold be 2.

More generally, let $\delta(k,l)$ be the number of mappings on k inputs with threshold l (including those with absolute threshold l). In computing $\delta(k,l)$ it will be convenient if we first calculate $\lambda(k,j,j')$, which is the number of mappings on k inputs with $l_0 = j$ and $l_1 = j'$. Note that by Lemma 4, $j + j'$ must be at least $k + 1$. We have the following theorem.

Theorem 4:

(i) $\lambda(k,j,j') = 2$ if $j + j' = k + 1$

(ii) $\lambda(k,j,j') = 2^{\binom{k}{j-1} + 2} - 6$ if $j + j' = k + 2$

(iii) $\lambda(k,j,j') = 4(2^{\binom{k}{j'-1}} - 1)(2^{\binom{k}{j-1}} - 1)2^{\sum_{i=k-j'+2}^{j-2} \binom{k}{i}}$

if $j + j' > k + 2$

Proof: Specifying j and j' fixes a common mapping value, say a, for all rows with j or more 0's and a common mapping value, say b, for all rows with $k - j'$ or fewer 0's. We must examine the possibilities for the remaining rows involving more than $k - j'$ 0's but fewer than j 0's. (i) Suppose that $j + j' = k + 1$. Then $k - j' = j - 1$ and there are no rows unaccounted for. If $a = 0$, then $b = 1$, and vice versa, hence $\lambda(k,j,j') = 2$.

(ii) Suppose that $j + j' = k + 2$. Then $k - j' = j - 2$ and thus only rows with $j - 1$ 0's must be considered. There are $\binom{k}{j-1}$ such rows. If $a = b = 0$, at least one of these rows must have mapping value 1. This may be done in $2\binom{k}{j-1} - 1$ ways. If a = b = 1, at least one of these rows must have mapping value 0, which again may be done in $2\binom{k}{j-1} - 1$ ways. If a = 1 and b = 0, then at least one row must have mapping value 1. This may be accomplished in $2\binom{k}{j-1} - 2$ ways. Similarly, this is so for a = 0 and b = 1. Combining these possibilities, we have

$$\lambda(k,j,j') = 2(2^{\binom{k}{j-1}} - 1) + 2(2^{\binom{k}{j-1}} - 2)$$

which after simplification yields (ii).

(iii) If $j + j' > k + 2$, then $k - j' < j - 2$. At least one row having $j - 1$ 0's must have mapping value $1 - a$, and at least one row having $k - j' + 1$ 0's $(k - j' + 1 < j - 1)$ must have mapping value $1 - b$. Rows involving more than $k - j' + 1$ but fewer than $j - 1$ 0's may be selected arbitrarily. Since a and b may each be 0 or 1, this allows four initial choices. Combining these possibilities yields (iii).

For any mapping m on k inputs with $l_0(m) \leq l_1(m)$ there is a symmetrically equivalent mapping (in terms of thresholds) m' given by

$$m'(x_1, x_2, \ldots, x_k) = m(1 - x_1, 1 - x_2, \ldots, 1 - x_k)$$

It is apparent that m' arises in the monotonic ordering of m by inverting the mapping value column and hence $l_1(m') = l_0(m) \leq l_1(m) = l_0(m')$. This symmetry implies that $\lambda(k,j,j') = \lambda(k,j',j)$ and thus finally enables us to calculate $\delta(k,l)$.

Theorem 5: $$\delta(k,l) = 2 \sum_{l'=l+1}^{k} \lambda(k,l,l') + \lambda(k,l,l')$$

where $\lambda(k,l,l') = 0$ if $l + l' < k + 1$.

Proof: The proof is contained in the discussion above.

4.6 Interrelationships Between Threshold, Forcibility, and Internal Homogeneity

In Sec. 4.4 we discussed the enumeration of Boolean transformations by internal homogeneity and forcibility. We now turn to the extension of these enumerations to include thresholds.

Given any mapping $m(\underset{\sim}{x})$ on k inputs, consider the mappings $m' = m(\underset{\sim}{1} - \underset{\sim}{x})$ (introduced in Sec. 4.5), $\bar{m} = 1 - m(\underset{\sim}{x})$ and $\bar{m}' - 1 - m(\underset{\sim}{1} - \underset{\sim}{x})$. It is easy to verify that the four mappings m, m', $\bar{m}$, and $\bar{m}'$ are equivalent with regard to internal homogeneity, forcibility, and thresholds. This equivalence class may consist of just two elements, m = m' or $m = \bar{m}'$ (obviously m cannot equal $\bar{m}$). If I is odd, four distinct mappings must arise; if $m = \bar{m}'$, then $I = 2^{k-1}$. Enumeration of the exact number of equivalence classes C(k) generated by grouping m, m', $\bar{m}$, $\bar{m}'$ is complicated. However, simple upper and lower bounds are readily available from the following theorem.

Theorem 6: $$\frac{1}{4}\sum_{i=2^{k-1}}^{2^k} N_k(i) \le C(k) \le \frac{1}{2}\sum_{m=2^{k-2}}^{2^{k-1}} N_k(2m) + \frac{1}{4}\sum_{m=2^{k-2}}^{2^{k-1}} N_k(2m+1)$$

where $N_k(i)$ is the number of mappings on k inputs with I = i.

Proof: It is easy to see that $N_k(i) = 2\binom{2^k}{i}$ for $2^{k-1} <$ $i \le 2^k$ and that $N_k(2^{k-1}) = \binom{2^k}{2^{k-1}}$

If I is odd, say 2m + 1, the number of equivalence classes generated is thus $(1/4)N_k(2m + 1)$. If I is even, say 2m, the number of classes generated is at least $(1/4)N_k(2m)$ and at most $(1/2)N_k(2m)$. Combining these results we have the theorem.

Similar equivalence classes developed for the special case of Boolean transformations with feedback have been considered in Walker and Aadryan (1971) and Gelfand and Walker (1977). They show that if k = 3, 88 equivalence classes arise. Theorem 6 provides bounds of 64 and 96.

When k is large, more than one equivalence class may have the same values for I, F, and *l*. This may occur even when k = 2, as the example in Table 4.3 indicates.

This example with discussion indicates that the three properties are clearly interrelated. Section 4.4 expanded on the strong relationship between internal homogeneity and forcibility; essentially, larger F must be accompanied by larger I. We shall eventually see that *l* is weakly related to I and inversely related to F.

We first jointly examine internal homogeneity and threshold. If for a mapping m on k inputs, $l(m) = 1$, then clearly $I = 2^k - 1$ (and in fact F = k).

Table 4.3 Two Mappings in Different Equivalence Classes with I = 2, F = 1, and *l* = 2

x_2	x_1	m_1	m_2
1	1	1	1
0	1	1	0
1	0	0	1
0	0	0	0

Conversely, if $I = 2^k - 1$, it is obvious that $l_0 + l_1 \leq k + 2$ and thus $1 \leq l \leq 1 + k/2$. Of the $2^k + 1$ mappings with $I = 2^k - 1$, how many have threshold l? The exact answer depends on whether k is odd or even. When k is odd, the number is $4\binom{k}{l-1}$. When k is even, the number is $4\binom{k}{l-1}$ if $1 \leq l \leq k/2$ and $2\binom{k}{k/2}$ if $l = k/2 + 1$.

We now enumerate $\tau(k,i,l)$, the number of mappings on k inputs having internal homogeneity i and threshold l. It will be convenient as in Theorem 5 to calculate $\tau(k,i,l_0,l_1)$ first. Also for convenience let

$$c = \max\left[\sum_{j=l_0}^{k}\binom{k}{j},\ \sum_{j=l_1}^{k}\binom{k}{j}\right],\quad d = \min\left[\sum_{j=l_0}^{k}\binom{k}{j},\ \sum_{j=l_1}^{k}\binom{k}{j}\right]$$

Recall also a and b as employed in Theorem 4.

We first note the following.

Lemma 5:

(i) If $a = b$, then $\tau(k,i,l_0,l_1) = 0$ if $i < c + d$.

(ii) If $a \neq b$, then $\tau(k,i,l_0,l_1) = 0$ if $i < c$ or $2^k - i < d$

Proof: Obvious from definitions.

To calculate $\tau(k,i,l_0,l_1)$, let $T(k,i,j,j')$ be the number of mappings on k inputs with $I = i$, $l_0 \leq j$ and $l_1 \leq j'$. If T is obtained, τ may be computed from T via a second-order difference, that is,

$$\tau = \Delta^2_{l_0,l_1}(T)$$

This notation means that

$$\tau(k,i,l_0,l_1) = T(k,i,l_0,l_1) - T(k,i,l_0-1,l_1) - T(k,i,l_0,l_1-1) + T(k,i,l_0-1,\ l_1-1)$$

Theorem 7 enables us to compute $T(k,i,l_0,l_1)$.

Theorem 7: Let

$$e_1 = \binom{2^k-(c+d)}{2^k-1}, \qquad e_2 = \binom{2^k-(c+d)}{i},$$

$$e_3 = \binom{2^k-(c+d)}{i-c}, \text{ and } \qquad e_4 = \binom{2^k-(c+d)}{i-d}$$

and define $\binom{x}{y} \equiv 0$ if $x < y$. Then

$$T(k,i,l_0,l_1) = \begin{cases} 2(e_1 + e_2 + e_3 + e_4) & \text{if } 2^{k-1} < i \leq 2^k \\ 2(e_1 + e_3) & \text{if } i = 2^{k-1} \end{cases}$$

Proof: Without loss of generality suppose that $l_0 \geq l_1$. Then in the monotonic ordering (by number of 0's), at least the last c input rows have map value fixed at a and at least the first c rows have map value fixed at b in order to ensure at most l_0 and l_1, respectively.

Suppose that i is the number of 1's in the table of mapping values. For the remaining 2^k - (c+d) rows, e_1 counts the number of ways we may ensure at most l_0 and l_1 with a = b = 1, e_2 indicates how many ways we may do this with a = b = 0, e_3 indicates how many ways with a = 1 and b = 0, and e_4 indicates how many ways with a = 0 and b = 1.

Since i may be the number of 0's in the table of values, we must multiply each e_i by 2. In the case that $i = 2^{k-1}$ we have $e_1 = e_2$, $e_3 = e_4$ and hence the expression for T is halved to avoid double counting.

Finally, we obtain $\tau(k,i,l)$.

Theorem 8:

$$\tau(k,i,l) = 2 \sum_{l'=l+1}^{k} \tau(k,i,l,l') + \tau(k,i,l',l)$$

Proof: It is obvious by symmetry considerations that

$$\tau(k,i,l,l') = \tau(k,i,l',l)$$

Our next objective is to examine forcibility and threshold. We have already observed that if for a mapping m $l(m) = 1$, then m is forcing on all inputs. If $l, 1 < l \leq (k+1)/2$ is an absolute threshold, consider any inputs, say the i-th. For any row with $x_i = 1$ and fewer than l inputs equal to 1 the mapping value will differ from that of a row with at least l inputs equal to 1. A similar argument holds when $x_i = 0$, so that no forcing inputs are possible.

The gist of the preceding paragraph may be stated as a lemma.

Lemma 6: The intersection of the set of forcible mappings and the set of mappings with an absolute threshold is the set of extended noncontractible mappings.

Proof: The proof is contained in the discussion above.

We now examine the more general situation. We show first that for any mapping having forcing inputs, both input values must have the same threshold state or $\max(l_0, l_1) = k$.

Theorem 9: Consider any mapping m on k inputs with differing threshold states and with $l_0(m) = j < k$ and $l_1(m) = j' < k$. Then m has no forcing inputs.

Proof: We show without loss of generality that the first input cannot be forcing. Consider the row with $x_1 = 1$ and $x_i = 0$, $i = 2, \ldots, k$ and the row with $x_i = 1$, $i = 1, \ldots, k - 1$ and $x_k = 0$. Since both j and j' are less than k, we must have $m(1,0,0,\ldots,1) \neq m(1,1,1,\ldots,1,0)$. Similarly, $m(0,1,1,\ldots,1) \neq$

m(0,0,0,...,0,1) and thus the first input cannot be forcing.

A partial converse to this result is available if we consider mapping having exactly f forcing inputs such that each forcing input has the same forcing state. For such a mapping the threshold cannot exceed k + 1 - f.

Theorem 10: Suppose that m is a mapping on k inputs which is forcing on exactly f of them and that each forcing input has the same forcing state. Then $l(m) \leq k + 1 - f$.

Proof: Without loss of generality we may assume that the first f inputs are forcing, that the forcing state is 1 and that the forced value is 1. If we examine any row having k + 1 - f or more 1's, at least one of them must be associated with one of the first f inputs and thus m = 1. Hence $l_1(m) \leq k + 1 - f$ and therefore $l(m) \leq k + 1 - f$.

We note two obvious corollaries.

Corollary 5: If f' of the f forcing inputs have forcing state 1 and f - f' having forcing state 0, then

$$l(m) \leq k + 1 - \max(f', f-f')$$

Corollary 6: If m is forcing on exactly k - 1 inputs and each forcing input has the same forcing state, then $l(m) = 2$.

Table 4.4 illustrates the need for all forcing inputs to have the same forcing state. Both mappings m_1 and m_2 have inputs 1 and 2 as forcing. However, m_1 satisfies the conditions of Theorem 10 (and Corollary 6) and has $l(m_1) = 2$. Mapping m_2 does not and has $l(m_2) = 3$ (in agreement with Corollary 5).

Table 4.4 Illustration of Theorem 10

x_3	x_2	x_1	m_1	m_2
1	1	1	1	1
0	1	1	1	1
1	0	1	1	1
0	0	1	1	1
1	1	0	1	0
0	1	0	1	0
1	0	0	0	1
0	0	0	0	1

We next turn to the calculation of $\sigma(k,f,l)$, the number of mappings on k inputs with exactly f forcing inputs and threshold l. The expression we develop is presented through a lemma and two theorems. The reason for its complexity is touched on in the discussion surrounding the previous theorem (i.e., forcing inputs need not have the same forcing state). As a result we must calculate $\rho(k,l_0,l_1,r,s)$, the number of maps on k inputs with thresholds l_0,l_1, respectively and r inputs having forcing state 1, s inputs having forcing state 0.

We first note:

Lemma 7: $\rho(k,l_0,l_1,r,s) = 0$ unless both $r \leq l_0$ and $s \leq l_1$.

Proof: Using the same argument as in Theorem 10, we see that $l_0 \leq k + 1 - s$, $l_1 < k + 1 - r$. Hence $s \leq k + 1 - l_0$, $r \leq k + 1 - l_1$. If $l_1 < s$, then $l_1 < k + 1 - l_0$, that is, $l_1 + l_0 < k + 1$, which is impossible. Similarly for $l_0 < r$.

Lemma 6 resolves the calculation of ρ when $l_0 + l_1 = k + 1$, that is, only $l = 1$ is possible and all inputs force. Thus $\rho(k,1,k,k,0) = \rho(k,k,1,0,k) = 2$. Furthermore, since

$$\lambda(k,l_0,l_1) = \sum_{\substack{s \leq k+1-l_0 \\ r \leq k+1-l_1}} \rho(k,l_0,l_1,r,s)$$

we may calculate $\rho(k,l_0,l_1,0,0)$ by subtraction if we have obtained $\rho(k,l_0,l_1,r,s)$ for any r, s with $\max(r,s) \geq 1$.

It is convenient to first calculate $R(k,l_0,l_1,r,s)$, which is the number of mappings on k inputs having thresholds l_0 and l_1 with *at least* the *first* r inputs having forcing state 1 and *at least* the *next* s inputs having forcing state 0. Lemma 8 calculates R, Theorem 11 shows how R' may be adjusted to yield ρ, and Theorem 12 obtains σ from ρ.

Lemma 8: Let

$$\eta(l_0,l_1,r) = \sum_{i=k-r-l_1+2}^{l_0-r-2} \binom{k}{i}$$

Case (i): $r \geq 1$, $s \geq 1$

If $l_0 + l_1 = k + 2$,

$$R(k,l_0,l_1,r,s) = 2(2^{\binom{k-(r+s)}{l_0-r-1}} - 1)$$

If $l_0 + l_1 > k + 2$,

$$R(,l_0,l_1,r,s) = 2(2^{\binom{k-(r+s)}{l_0-r-1}}-1)\ (2^{\binom{k-(r+s)}{l_1-s-1}}-1)2^{\eta(l_0,l_1,r)}$$

Case (ii): $r \geq 1$, $s = 0$

If $l_0 = k, l_0 + l_1 = k + 2$,

$$R(k,k,2,r,0) = 2^{k-r+2} - 6$$

If $l_0 = k, l_0 + l_1 > k + 2$,

$$R(k,k,l_1,r,0) = 4(2^{k-r} - 1)(2^{\binom{k-r}{l_1-1}} - 1)2^{\eta(k,l_1,r)}$$

If $l_0 < k$, same expressions as in case (i).

Case (iii):

If $l_1 = k, l_0, l_1 = k + 2$,

$$R(k,2,k,0,s) = 2^{k-s+2} - 6$$

If $l_1 = k, l_0 + l_1 > k + 2$,

$$R(k,l_0,k,0,s) = 4(2^{\binom{k-s}{l_0-1}} - 1)(2^{k-s} - 1)2^{\eta(l_0,k,s)}$$

If $l_0 < k$, same expressions as in case (i).

Proof: The proof of this lemma is quite similar to that of Theorem 4. We prove only cases (i) and (ii) since case (iii) is symmetrically equivalent to case (ii).

(i) $r \geq 1$, $s \geq 1$ implies in the lexicographic order that the first $2^k - 2^{k-r}$ rows (i.e., those which have a 1 for any of the first r inputs) are determined at, say, $\underline{a}$. Also, the last $2^{k-r} - 2^{k-r-s}$ rows must as well be determined at this $\underline{a}$ (i.e., the forced value is unique and these are the remaining rows having a 0 in at least one of the next s inputs). Note that since the first and last rows have map value $\underline{a}$, both input values must have the same threshold state, $\underline{a}$. Consider the undetermined rows. All have $x_1 = x_2 = \ldots = x_r = 0$, $x_{r+1} = x_{r+2} = \ldots = x_{r+s} = 1$. If $l_0 = r$ or $l_1 = s$, all of these rows must also be determined to have map value $\underline{a}$ and thus the trivial map results. Hence we take $r < l_0$, $s < l_1$. To have threshold l_0 and l_1, for the remaining $k - (r+s)$ inputs whenever $l_0 - r$ or more are 0 or whenever $l_1 - s$ or more are 1, the map value must again be $\underline{a}$. To ensure exactly l_0, l_1 we must look among these $k - (r+s)$ inputs at rows with exactly $l_0 - r - 1$ of them at 0 and at rows with exactly $l_1 - s - 1$ of them at 1 [i.e., $k - (r+s) - (l_1 - s - 1) = k - r - l_1 + 1$ of them at 0].

If $l_0 + l_1 = k + 2$, then $k - r - l_0 + 1 = l_0 - r - 1$ and for at least one of these $\binom{k - (r+s)}{l_0 - r - 1}$ rows the map value must be $1 - a$. Since $\underline{a}$ may be chosen in two ways, the expression for $l_0 + l_1 = k + 2$ follows.

If $l_0 + l_1 > k + 2$, then $k - r - l_1 + 1 < l_0 - r - 1$ and the second part of (i) follows as in (iii) of Theorem 4 except that again $\underline{a}$ can be chosen in two ways and b must be $1 - \underline{a}$.

(ii) $r \geq 1$, $s = 0$ implies again that the first $2^k - 2^{k-r}$ rows are determined at, say, $\underline{a}$. However, with $s = 0$ the map value for the last row is not fixed. From Theorem 9 with $1 \leq r \leq l_0$ (Lemma 7), if the threshold states differ, l_0 must equal k and the last rows will have map value $1 - \underline{a}$. Hence again as in Theorem 4, if $l_0 = k$ we may have $\underline{a} \neq b$, but if $l_0 < k$, we must have $\underline{a} = b$. For the former the expressions mimic cases (ii) and (iii) of Theorem 4, while for the latter the same argument as in case (i) of this theorem is appropriate with $s = 0$. Hence the expressions in (ii) follow and we are done.

Note that the completion of the rows in cases (i) and (ii) in order to fix l_0 and l_1 may result in more than just the first r inputs being forcing with forcing state 1 and more than just the next s inputs being forcing with forcing state 0.

Theorem 11:

$$\rho(k,l_0,l_1,r,s) = \frac{k!}{r!s!(k-(r+s))!} \sum_{\substack{0\le j,j' \\ j+j'\le k-(r+s)}} (-1)^{j+j'}$$

$$\times \frac{(k-(r+s))!}{j!j'!(k-(r+s)-j-j')!} R(k,l_0,l_1,r+j,s+j')$$

Proof: The summation adjusts R to the number of ways in which exactly the first r inputs have forcing state 1 and exactly the next s inputs have forcing state 0. The form may be established from a straightforward counting argument using symmetry in the selection of the additional j + j' forcing inputs. The details are omitted. The factorial coefficient allows the adjustment from the first r and next s inputs to an arbitrary choice of r and s from the total of k inputs.

We finally have

Theorem 12:

$$\sigma(k,f,l) = 2 \sum_{l'=l+1} \sum_{r+s=f} \rho(k,l,l',r,s)$$

$$+ \sum_{r+s=f} \rho(k,l,l',r,s)$$

Proof: In order to have exactly f forcing inputs, r + s must equal f. Since it is apparent that

$$\sum_{r+s=f} \rho(k,l_0,l_1,r,s) = \sum_{r+s=f} \rho(k,l_1,l_0,r,s)$$

the conclusion follows.

Considering the difficulties involved in achieving Theorems 8 and 12, we do not undertake theoretical enumeration of maps by *l* and I and F. Instead, we offer Table 4.5, which presents such an explicit enumeration

Table 4.5 Enumeration of Mappings by Number of Forcing Inputs (f), Internal Homogeneity (i), and Threshold (*l*), for Number of Inputs k = 2, 3, 4

k = 2

l = 2

f \ i	2	3	4	
0	2	-	-	2
1	4	-	-	4
2	-	4	-	4
	6	4	-	10

l = 1

f \ i	2	3	4	
0	-	-	-	-
1	-	-	-	-
2	-	4	-	4
	-	4	-	4

l = 0

f \ i	2	3	4	
0	-	-	-	-
1	-	-	-	-
2	-	-	2	2
	-	-	2	2

k = 3

l = 3

f \ i	4	5	6	7	8	
0	62	60	8	-	-	130
1	6	36	12	-	-	54
2	-	-	12	-	-	12
3	-	-	-	-	-	-
	68	96	32	-	-	196

Table 4.5 (Continued)

		i = 4	5	6	7	8	
l = 2	f 0	2	4	-	-	-	6
	1	-	12	12	-	-	24
	2	-	-	12	-	-	12
	3	-	-	-	12	-	12
		2	16	24	12	-	54

		i = 4	5	6	7	8	
l = 1	f 0	-	-	-	-	-	-
	1	-	-	-	-	-	-
	2	-	-	-	-	-	-
	3	-	-	-	4	-	4
		-	-	-	4	-	4

		i = 4	5	6	7	8	
l = 0	f 0	-	-	-	-	-	-
	1	-	-	-	-	-	-
	2	-	-	-	-	-	-
	3	-	-	-	-	2	2
		-	-	-	-	2	2

Table 4.5 (Continued)

k = 4

l = 4	f	i = 8	9	10	11	12	13	14	15	16	
	0	12,242	21,288	13,814	6,340	1,830	260	10	-	-	55,784
	1	8	112	336	560	496	192	16	-	-	1,720
	2	-	-	-	-	24	48	24	-	-	96
	3	-	-	-	-	-	-	-	-	-	-
	4	-	-	-	-	-	-	-	-	-	-
		12,250	21,400	14,150	6,900	2,350	500	50	-	-	57,600

l = 3	f	i = 8	9	10	11	12	13	14	15	16	
	0	620	1,464	1,754	1,496	734	156	6	-	-	6,230
	1	-	16	112	336	512	304	48	-	-	1,328
	2	-	-	-	-	24	12	48	-	-	192
	3	-	-	-	-	-	-	48	-	-	48
	4	-	-	-	-	-	-	-	12	-	12
		620	1,480	1,866	1,832	1,270	580	150	12	-	7,810

l = 2	f	i = 8	9	10	11	12	13	14	15	16	
	0	-	-	-	4	4	-	-	-	-	8
	1	-	-	-	-	16	16	-	-	-	32
	2	-	-	-	-	-	24	24	-	-	48
	3	-	-	-	-	-	-	16	-	-	16
	4	-	-	-	-	-	-	-	16	-	16
		-	-	-	4	20	40	40	16	-	120

Table 4.5 (Continued)

$l = 1$	f \ i	8	9	10	11	12	13	14	15	16	
	0	-	-	-	-	-	-	-	-	-	-
	1	-	-	-	-	-	-	-	-	-	-
	2	-	-	-	-	-	-	-	-	-	-
	3	-	-	-	-	-	-	-	-	-	-
	4	-	-	-	-	-	-	-	4	-	4
		-	-	-	-	-	-	-	4	-	4

$l = 0$	f \ i	8	9	10	11	12	13	14	15	16	
	0	-	-	-	-	-	-	-	-	-	-
	1	-	-	-	-	-	-	-	-	-	-
	2	-	-	-	-	-	-	-	-	-	-
	3	-	-	-	-	-	-	-	-	-	-
	4	-	-	-	-	-	-	-	-	2	2
		-	-	-	-	-	-	-	-	2	2

for k = 2, 3, and 4. The theoretical results for *l* and I (Theorem 8) may be verified by summing the tables over F, similarly for *l* and F (Theorem 12) by summing over I. The tables, particularly at k = 3 and 4, reveal a weak relation between *l* and I and a rather strong inverse relation between *l* and F.

In fact, for general k the relationship between *l* and I must continue to be weak. We recall, using the monotonic ordering for mappings, that only an upper and a lower set of input rows are determined. The remaining middle rows are all free to assume either map value. Hence, if k is large and *l* is not too small, I is only weakly controlled by the specification of *l*. In the situation where *l* is very small or is an absolute threshold, this will not be the case but such mappings are sparse in the overall collection of possible mappings. For general k the strong inverse relationship between *l* and F also persists by reference to Theorem 10 and Corollary 5. More precisely, decreasing threshold has the effect of increasing the proportion of maps in the given threshold class that have one or more forcible inputs. For example, with k = 4 maps, that proportion is 0.03 at $l = 4$, 0.20 at $l = 3$, 0.93 at $l = 2$, and 1.0 at $l = 1$ and $l = 0$. From Table 4.5 it can also be seen that within threshold classes the direct relationship between internal homogeneity and forcibility noted in Sec. 4.4 still obtains. This is a nontrivial finding in that contrary outcomes are conceivable, particularly for $l = 2$. Extended threshold could vary the way in which internal homogeneity and forcibility are related. We find that it does not. This suggests the existence of some mechanism common to the three measures.

4.7 Other Behavioral Variables

Two less frequently discussed behavioral variables are discussed in this section. This material is drawn from Gelfand and Walker (1982). These are the density of 1's (hence of 0's) in net states and the (Hamming) distance between net states. The density of 1's can be used to characterize net states, that is, allowing the net sufficient time to go into cycle, it can be used to typify cyclic states. The Hamming distance measures the number of net elements which have changed state as the net moves from one time point to the next. When expressed as a fraction of the total number of net elements, it describes net "activity" (see, e.g., Walker and Ashby, 1966).

For a binary switching net of N elements with input connectance k, assume a completely random (i.e., equally likely) assignment of inputs to each element. Let

N_t = number of elements in state 1 at time t

$D_t = N_t/N$ = density of 1's at time t

M_t = number of elements which changed state from time t - 1 to time t;

$C_t = M_t/N$ = density of changes from t - 1 to t.

Note the following:

1. t = 0 denotes the starting state of the net; that is, N_0 is the initial number of 1's. M_0 is not defined.
2. M_t is the Hamming distance between the state of the net at t - 1 and the state at t.

We seek the distribution of N_t (equivalently D_t) and of M_t (equivalently C_t) assuming that each net element is assigned a Boolean transformation drawn:

1. Equally likely from each of the 2^{2^k} possibilities
2. Randomly from a subset of mappings determined by a specified level of one of the forms of functional control described in the preceding sections

For a Boolean transformation thus selected, let B = number of 1's (this, of course, is the inernal homogeneity of the transformation, but under the development in Sec. 4.7.1 it is useful to use B to distinguish the actual or realized number of 1's from a "controlled" or predetermined number of 1's) and α = P (output value changes given input vector changed).

In considering the distributions of the variables N_t and M_t it is important to understand that our models do not yield a "Markovian" system. As a simple illustration, let N = 2, k = 1. Then in the equally likely, fully random situation [(i) above] it is easy to verify that

$P(N_2 = 0 \text{ given } N_1 = 2, N_0 = 0) = 1/4$
$P(N_2 = 0 \text{ given } N_1 = 2, N_0 = 1) = 1/16$
$P(N_2 = 0 \text{ given } N_1 = 2, N_0 = 2) = 0$

that is, one must know N_0 to describe the conditional distribution $N_2|N_1$. More generally the conditional probability $P(N_2 = n'|N_{t-1} = n)$ is not well defined without specifying N_0 and in fact $N_1, N_2, \ldots N_{t-2}$ [e.g., $P(N_t = N|N_{t-1} = N, N_{t-2} = N) = 1$].

However, suppose that N is large and N_0 is not extreme. Then, by virtue of the complete randomness of the interconnectance structure, P(element s is in state 1 at time $t|N_{t-1} = n)$, $t \geq 2$, will be approximately constant over such N_0 and in fact over $N_1, N_2, \ldots, N_{t-2}$. Hence we make the assumption that this conditional probability is essentially well defined and denote it by $\theta(n)$. Similarly, we take the conditional probability P(element s changes state from time t - 1 to $t|M_{t-1} = m)$,

$t \geq 2$, as essentially well defined and denote it by $\phi(m)$. Thus $\phi(m)$ becomes

$$\phi(m) = \alpha\left[1 - \binom{N-m}{k}\Big/\binom{N}{k}\right] \qquad (4.86)$$

These assumptions enable immediate approximations to the expected values of the random variables of interest in the form of recurrence relationships:

$$\begin{aligned} E(N_t) &= E\,E(N_t|N_{t-1}) = E[N\theta(N_{t-1})] = N\,E[\theta(N_{t-1})] \\ E(D_t) &= E[\theta(ND_{t-1})] \\ E(M_t) &= N\,E[\phi(M_{t-1})] \\ E(C_t) &= E\,\phi(NC_{t-1}) \end{aligned} \qquad (4.87)$$

with $E(N_1) = N\,\theta(N_0)$ and $E(M_1) = N\,\gamma(N\theta)$ where $\gamma(N_0)$ is the probability that state s changes from its initial state given N_0 1's in the initial state. Note that the expectations in (4.87) are exact given assumptions and are approximately true. Hence we use = rather than ≈ there and in what follows in expressing theoretically correct relationships given assumptions.

Two further approximating assumptions are noteworthy. When N is large, the complete randomness of the interconnectance structure suggests that

1. Given N_{t-1}, element states at time t are approximately independent of each other.
2. Given M_{t-1}, element changes from time t - 1 to t are approximately independent of each other. Hence

$N_t|N_{t-1} = n$ will be approximately distributed as binomial $[N,\theta(n)]$, and $M_t|M_{t-1} = m$ will be approximately distributed as binomial $[N,\phi(m)]$ for n,m not extreme. (Henceforth, we shorten binomial to Bi.) Note that N_1 is exactly distributed as $Bi(N,\theta(N_0))$ and M_1 is exactly distributed as $Bi(N,\gamma(N_0))$.

Hence with $t \geq 2$

$$P(N_t = n') = \sum_{n=0}^{N} \binom{N}{n'} [\theta(n)]^{n'} [1-\theta(n)]^{N-n'} P(N_{t-1} = n)$$

$$P(M_t = m') = \sum_{m=0}^{N} \binom{N}{m'} [\phi(m)]^{m'} [1-\phi(m)]^{N-m'} P(M_{t-1} = m) \tag{4.88}$$

The expressions in (4.88) are less useful than they might appear. Although in principle we can recursively build up approximations to the distributions of N_t, M_t from N_1, M_1, respectively, the expressions quickly become intractable. Of course, if $\theta(n) \equiv \theta$ ($\phi(m) \equiv \phi$), then $N_t \sim \text{Bi}(N, \theta)$, ($M_t \sim \text{Bi}(N, \phi)$).

An effort to find an equilibrium (stationary) distribution (letting $t \to \infty$) in (4.88) produces in either case a system of equations of the form (familiar in Markov chain theory)

$$p = Qp \iff (I - Q)p = 0 \tag{4.89}$$

where p is an $(N + 1) \times 1$ vector of probabilities, Q is an $(N + 1) \times (N + 1)$ matrix of binomial probabilities whose columns sum to 1, and I is an $(N + 1) \times (N + 1)$ identity matrix. But $I - Q$ is not of full rank since its columns sum to 0, and in fact for the nontrivial functions θ, ϕ to be obtained in the next section, the rank of $I - Q$ is n. Hence there will be a unique stationary distribution which solves (4.89), but no general expression for it is available.

The approximate conditional distribution above can also be used to approximate higher-order moments of N_t and M_t.

4.7.1 Four Cases of Interest

We apply the results just obtained to four particular cases. As case I we take the fully random net with input connectance k. For cases II, III, and IV, we make some simplifying assumptions.

We simplify IH to the *number of* 1 *responses*. In asserting that a Boolean transformation has IH equal to i, we mean only that I of the 2^k input vectors have been selected at random and 1 assigned as output. A distribution guides the assignment of 0's and 1's to the remaining 2^k - i input vectors. Thus IH is at least i and we may speak of $P(IH = j|IH \geq i)$. As case II we take nets with input connectance k such that each Boolean transformation has IH $\geq$ i.

For forcibility we simplify to *forcing state* 1 and *forced value* 1. In asserting that a Boolean transformation has f forcing inputs, we mean only that (without loss of generality) the first f inputs are such that if any of them are at 1, in an input vector the output value will be 1. Again a distribution guides the assignment of 0's and 1's to the remaining input vectors. Thus if F denotes the number of forcing inputs, F is at least f and we may speak of $P(F = j|F \geq f)$. As case III we take nets with input connectance k such that each Boolean transformation has F $\geq$ f.

For extended threshold, we simplify to *l = the smallest number of* 1 *inputs which ensure a* 1 *output*. In asserting that a Boolean transformation has threshold *l*, we mean only that for any input vector having *l* or more 1's, the output value will be 1. Again a distribution guides the assignment of 0's and 1's to the remaining input vectors. Thus, if L denotes the actual threshold, L is at most *l* and we may speak of $P(L = j|L \leq l)$. As case IV we take nets with input connectance k such that each Boolean transformation has L $\leq$ *l*.

For the reader who questions these simplifications, we note that in many applications the state 1 is of greater interest, typically indicating that the element is "on" or "active." When inputs to an element are active, we would probably be concerned with whether the element will be activated. Moreover, after examining results in the sequel, the reader will recognize that, where it makes sense, ready extension to the more general definition can be given.

For each of these four cases, we now obtain the distribution of B; we calculate α, θ, and ϕ and where possible characterize the resultant sequences of expectations for the variables N_t, D_t, M_t, and C_t.

Case I. In this case the assignment of an output value to an input vector is made equally likely between 0 and 1. Hence $B \sim Bi(2k,\frac{1}{2})$, $E(B) = 2^{k-1}$, and $\alpha = \frac{1}{2}$. Thus $\theta(n) = \frac{1}{2}$ and from (4.86), $\phi(m) = \frac{1}{2}(1 - (N-m)_k/(N)_k)$. $\gamma(N_0)$ is clearly $\frac{1}{2}$. Therefore, $E(N_t) = N/2$, $E(D_t) = \frac{1}{2}$. If $k = 1$, $\phi(m) = m/2N$, whence $E(M_t) = N2^{-t}$, $E(C_t) = 2^{-t}$. Hence in a large $k = 1$ net, the density of changes (i.e., net activity) will be near 0 fairly quickly. If $k = N$ (i.e., a fully connected net), $\phi(m) = \frac{1}{2}$ and $E(M_t) = N/2$, $E(C_t) = \frac{1}{2}$.

For general k the recurrence relationships for $E(M_t)$ and $E(C_t)$ quickly become intractable. The fact that, for arbitrary k, $\phi(m)$ is concave increasing on [1,N] enables simple bounds on these sequences.

Case II. As noted earlier, imposing $IH \geq i$ leaves open the specification of the distribution which guides the assignment of 0's and 1's to the remaining $2^k - i$ input vectors. Two simple possibilities are the binomial and the truncated geometric. Let $p_b \equiv P(B = b|IH \geq i)$, $b = i, i + 1, \ldots, 2k$. In the former $B = U + i$ where $U \sim Bi(2^k - i, \frac{1}{2})$ and $E(B) = 2^{k-1} + i/2$. If $i = \delta 2^k$,

$0 < \delta < 1$, then $2^{-k}E(B)$ tends to $(1 + \delta)/2$ as k grows large. In the latter $p_{b+1} = \frac{1}{2}p_b$, whence

$$p_b = 2^{2^k - b}/(2^{2^k - i+1} - 1)$$

and

$$E(B) = \frac{(i+1)2^{2^k - i+1} - (2^k + 2)}{2^{2 - i+1} - 1}$$

If $i = \delta 2^k$, then $2^{-k}E(B)$ decreases to δ as k grows large. Under any distribution, $\theta(n) = 2^{-k}E(B)$, whence $E(N_t) = N\theta$, $E(D_t) = \theta$. For α we have

$$\alpha = \sum_{b=1}^{2^k} P(\text{output value changes given input vector changed}|b)p_b$$

$$= \sum_{b=1}^{2^k} \left(\frac{b}{2^k}\ \frac{2^k-b}{2^k-1} + \frac{2^k-b}{2^k} \cdot \frac{b}{2^k-1}\right) p_b$$

Obviously, when $i = 2^n$, $\alpha = 0$. Under the binomial model, we obtain

$$\alpha = \frac{2i(2^k-i) + (2^k-i)(2^k-i-1)}{2^{k+1}(2^k-1)} \tag{4.90}$$

If $i = \delta 2^k$, α tends to $(1-\delta^2)/2$ as k grows large. Under the truncated geometric model, we obtain ($k \geq 2$, $i < 2^k$)

$$\alpha = \frac{4(2^k+3) + 2^{2^k - i+1}[(2^k-i-1)(i+1) - 2]}{2^k(2^k-1)(2^{2^k - i+1} - 1)}$$

(At $k = 1$, from (5), $\alpha = 2/7$ at $i = 0$, $\alpha = 2/3$ at $i = 1$.) If $i = \delta 2^k$, α tends to $2\delta(1 - \delta)$ as k grows large. In any event, $\phi(m)$ is as in (4.86), and thus if $k = 1$, $\phi(m) = m/N$, so that, using (4.87)

$$E(M_t) = \alpha^{t-1}E(M(1)) = N\alpha^{t-1}\gamma(N_0) \quad (\to 0 \text{ as } t \to \infty)$$
$$E(C_t) = \alpha^{t-1}E(C(1)) = \alpha^{t-1}\gamma(N_0) \quad (\to 0 \text{ as } t \to \infty) \tag{4.91}$$

where $\gamma(N_0 = N^{-1}\ 2^{-k}[(N - N_0)E(B) + N_0(2^k - E(B)]$.
If $k = N$, $\theta/(m) = \alpha$ and $E(M_t) = N\alpha$, $E(C_t) = \alpha$.

Case III. As noted earlier, imposing $F \geq f$ fixes the output value at 1 for the $2^k - 2^{k-f}$ input vectors having at least one of the first f inputs at 1. We consider a binomial model in assigning output values to the remaining 2^{k-f} rows. Let π denote the probability that the output value is 1 for an input vector with no forcing input at 1. We have $B = U + 2^k - 2^{k-f}$, where $U \sim Bi(2^{k-f},\pi)$ and $E(B) = 2^k - (1-\pi)2^{k-f}$. With n of the N elements at 1 and with f forcing inputs, we have

$$\theta(n) = 1\ [1 - \binom{N-n}{f}\Big/\binom{N}{f}] + \pi\binom{N-n}{f}\Big/\binom{N}{f}$$
$$= 1 - (1 - \pi)\binom{N-n}{f}\Big/\binom{N}{f}$$

from which we see that $\theta(n)$ is concave, increasing on $[0,N]$. At $f = 1$, $\theta(n) = \pi + (1 - \pi)n/N$, so that

$$E(N_t) = N - (1 - \pi)^t (N - N_0)$$
$$E(D_t) = 1 - (1 - \pi)^t \left(\frac{N - N_0}{N}\right) \tag{4.92}$$

Note that if $\pi = 0$, $E(D_t) = N_0/N$. Otherwise, $E(D_t) \to 1$ as $t \to \infty$. Moreover, if $f > 1$, $E(D_t) \to 1$ more quickly. We may thus conclude that if there is at least one forcing input with forcing state 1, the density of 1's in a large net will tend to 1.

A computation similar to (4.90) reveals that $(f \geq 1)$

$$\alpha = \frac{(1 - \pi)(2^k - 2^{k-f}) + \pi(1 - \pi)(2^{k-f} - 1)}{2^{f-1}(2^k - 1)}$$

As k grows large, α tends to $2^{-(f-1)}[(1-\pi) - 2^{-f}(1-\pi)^2]$. When f = 1 this reduces to $(1 - \pi^2)/2$.

Again, $\phi(m)$ is as in (4.86) and if k = 1, $\phi(m) = m/N$, so that $E(M_t)$, $E(C_t)$ are as in (4.91), where

$$\gamma(N_0) = \frac{N-N_0}{N} + (1 - \pi)(2\frac{N_0}{N} - 1) \binom{N-N_0}{f} \Big/ \binom{N}{f}$$

Again if k = N, $\phi(m) = \alpha$ and $E(M_t) = N\alpha$, $E(C_t) = \alpha$.

Case IV. As noted earlier, imposing $L \leq l$ fixes the output value at 1 for all $w_l = \sum_{j=l}^{k} \binom{k}{j}$ input vectors having at least l 1's. We again consider a binomial model in assigning output values to the remaining $2k - w_l$ input vectors. Let π denote the probability that the output value is 1 for an input vector with fewer than l inputs at 1. We have $B = U + w_l$ where $U \sim Bi(2^k - w_l, \pi)$ and $E(B) = 2k\pi + (1 - \pi)w_l$. Moreover

$$\theta(n) = v_l + \pi(1 - v_l)$$

where

$$v_l = \sum_{j=l}^{k} \binom{n}{j} \binom{N-n}{k-j} \Big/ \binom{N}{k}$$

When l = 1, it is straightforward to show that $E(D_t) \to 1$ as $t \to \infty$. Only when k = 1 are expressions simple, for example, $\theta(n) = \pi + (1-\pi)n/N$ and $E(N_t)$, $E(D_t)$ are as in (4.92).

A computation similar to (4.90) reveals that $(l \leq k)$

$$\alpha = \frac{(1-\pi)w_l(2^k-w_l) + (2^k-w_l)(2^k-w_l-1)\pi(1-\pi)}{2^{k-1}(2^k-1)}$$

Again, $\phi(m)$ is as in (4.86) and if k = 1, $E(M_t)$ are as in (4.91), where

$$\gamma(N_0) = \pi(1-v_1) + \binom{N-N_0}{N} v_1$$

Again if $k = N$, $\phi(m) = \alpha$ and $E(M_t) = N\alpha$, $E(C_t) = \alpha$.

The results of a simulation study involving these four cases are taken up in Sec. 5.3.4. Of particular interest are the findings when $k > 1$, but small.

References

Babcock, A. K. (1976). Logical probability models and representation theorems on the stable dynamics of the Genetic Net. Doctoral Dissertation, University of New York at Buffalo, 1976.

Cull, P. (1971). Linear analysis of switching nets, *Kybernetik*, *8*, 31-39.

Cull, P. (1978). A matrix algebra for neural nets. In G. J. Klir (Ed.), *Applied General Systems Research*. New York: Plenum Press, pp. 563-573.

Ferguson, T. (1967). *Mathematical Statistics: A Decision Theoretic Approach*. New York: Academic Press.

Folkert, J. E. (1955). The distribution of the number of components of a random mapping function. Doctoral Dissertation, Michigan State University.

Gelfand, Alan E. (1982). A behavioral summary for completely random nets. *Bulletin of Mathematical Biology*, *44*, 309-320.

Gelfand, A. E. and Walker, C. C. (1977). The distribution of cycle lengths in a class of abstract systems. *International Journal of General Systems*, *3*, 39-46.

Gelfand, A. E. and Walker, C. C. (1979). Management strategies in fixed-structure models of complex organizations II. Techical Report 250, Department of Statistics, Stanford University.

Gelfand, A. E. and Walker, C. C. (1982). On the character of and distance between states in a binary switching net. *Biological Cybernetics*, *43*, 79-86.

Gontcharoff, W. (1944). On the field of combinatory analysis. *Bulletin de l'Académie des Sciences de l'U.R.S.S., Série Mathématique, 8*, 1-48.

Harris, B. (1960). Probability distributions related to random mappings. *Annals of Mathematical Statistics, 31*, 1045-1062.

Johnson, N. and Kotz, S. (1969). *Discrete Distributions*. Boston: Houghton-Mifflin.

Katz, L. (1955). Probability of indecomposability of a random mapping function. *Annals of Mathematical Statistics, 26*, 512-517.

Kauffman, S. A. (1969). Metabolic stability and epigenesis in randomly constructed genetic nets. *Journal of Theoretical Biology, 22*, 437-467.

Kauffman, S. A. (1970). The organization of cellular genetic control systems. *Mathematics in the Life Sciences, 3*,63-116.

McCulloch, W. S. and Pitts, W. H. (1943). A logical calculus of the ideas immanent in nervous activity. *Bulletin of Mathematical Biophysics, 5*, 115-133.

Newman, S. A. and Rice, S. A. (1971) Model for constraint and control in biochemical networks. *Proceedings of the National Academy of Sciences, 68*, 92-96.

Riordan, J. (1958). *An Introduction to Combinatorial Analysis*. New York: Wiley.

Rosen, R. (1958). A relational theory of biological systems. *Bulletin of Mathematical Biology, 20*, 245-260.

Rubin, H. and Sitgreaves, R. (1954). Probability distributions related to random transformations on a finite set. Technical Report 19A, Applied Mathematical and Statistical Laboratory, Stanford University.

Sherlock, R. (1979). Analysis of the behavior of Kaufman binary networks--I. State space descriptions and the distribution of limit cycle lengths. *Bulletin of Mathematical Biology, 41*, 687-705.

Walker, C. C. (1965). A study of a family of complex systems: An approach to the investigation of organisms' behavior. Technical Report 6, AF Grant 7-64, June 1965, Electrical Engineering Research Laboratory, University of Illinois, Urbana.

Walker, C. C. and Aadryan, A. A. (1971). Amount of computation preceding externally detectable steady state behavior in a class of complex systems. *Journal of Biomedical Computing, 2,* 85-94.

Walker, C. C. and Ashby, W. R. (1966). On temporal characteristics of behavior in certain complex systems. *Kybernetik, 3,* 100-108.

Walker, C. C. and Gelfand, A. E. (1977). Management strategies in fixed-structure models of complex organizations. Technical Report 243, Department of Statistics, Stanford University.

Walker C. C. and Gelfand, A. E. (1979). A system theoretic approach to the management of complex organizations: Management by exception, priority and input span in a class of fixed-structure models. *Behavioral Science, 24,* 112-120.

5

Empirical Studies of Ensembles of Simple Switching Nets

5.1 Introduction

This chapter discusses empirically derived details of the behavior of ensembles of simple switching nets. The presentation is divided into sections describing two main categories of nets: functionally homogeneous, where all net elements compute the same logical function, and functionally heterogeneous, where each element's mapping is selected from some specified ensemble of functions.

Why study these nets empirically? As pointed out earlier, given the present state of analytic technique, the behavior of many interesting classes of nets is most conveniently determined by direct observation, that is, by simulation. Even where the behavior of interest is addressed by analysis, practical difficulties often intrude, such as having to handle actual matrices that are $2^N \times 2^N$, a formidable task where N is 20 or more.

Although it is relatively easy through simulation to observe the behavior of nets that would be difficult to deal with analytically, there are unavoidable limits on what can be achieved empirically. Perhaps the main embarrassment of empirical methods here has to do with the vastness of these nets' behavior or cycle spaces in comparison with the size of the behavioral sample. If the aim is to determine run-in or cycle lengths, the fundamental empirical procedure is: "Build" the net in question, simulate a trajectory of net states, and directly examine these states for a cycle. If we call the number of states in the observed trajectory sequence the "observational aperture," a, it follows that the largest cycle observable has length $a - 1$, since there must be at least one state repeated. The largest detectable run-in is $a - 2$ states long, since the smallest cycle has length 1, and there must be one repetition of it. The longest disclosure then has length $a - 1$. Thus we will not be able to detect *any* cycle that is a or more states in length. Nor will we be able to detect *any* cycle that happens to be $a - 1$ or more states away from the first state in the trajectory. Nor will we be able to detect any cycle if its length plus the distance from the initial state is a or more. That is, disclosure lengths stand as a barrier to the empirical observation of run-in and cycle lengths. Our ability to surmount that barrier depends on the cost of using apertures sufficiently large and search techniques sufficiently fast.

5.1.1 Aperture Size Versus Behavior Space Size

It is instructive to compare aperture size to typical behavior space sizes. A fairly large aperture by present-day standards might be about 10,000 states. By comparison, the aperture used for most runs in Walker (1965), using the conveniently available technology of 1963, was 500 states. The largest aperture that could be accommodated using the IBM 7094 computer, on which the program was run, was 5180 states for nets of size 100.

To compare aperture and behavior space, consider the 2^N states of the behavior space to be distributed evenly over a flat disk at a density which puts 10,000 states into a circle with a diameter of 1 inch. Although this two-dimensional packing of behavior sequences which are perhaps more properly one-dimensional may be awkward, it allows us in some sense to think of our aperture as roughly the size of a 25-cent piece. At this scale, a little arithmetic will show that the behavior space of a 100-element net has a diameter of approximately 178 million miles--roughly the diameter of the earth's orbit around the sun. Empirical investigation of the behavior of the 100-element net is then somewhat like throwing handfuls of coins over a carpet millions of miles in extent, and trying to describe the appearance of the carpet from what can be seen under the coins. It is clear that this procedure will reveal, if any patterns at all, only those that are very fine-grained with respect to what the carpet could display.

Now suppose that the net is increased tenfold to 1000 elements. What happens to the size of the behavior space? The 2^{1000}-net states, on the scale at which the aperture is 1 inch across, now require a disk whose diameter is about 3×10^{131} light-years. This is a span incomprehensibly larger than the diameter of the known universe. The effect of net size on behavior space size is, to say the least, dramatic. Since the aperture is relatively bounded, empirical study of even moderately large switching nets is limited. We can directly examine only a numerically insignificant portion of the full behavior space. Under such conditions, it would be no surprise to find no cycles at all under our "coins." If cycles should be found, this would indicate a sharp localization of behavior.

5.1.2 Initial States as a "Technical" Ensemble

It is inescapable that the process of simulating behavior requires an initial state to be chosen for each behavior sequence observed. This, in turn, means that some criterion be available for making the choices of initial state. If the modeling context specifies that uncertainty exists with respect to initial states, then, in our view, the use of an appropriate ensemble of initial states is indicated.

The connection of the foregoing observation with the present chapter lies in the fact that most empirical studies of switching net ensembles have not set out their modeling contexts well enough to specify unarguably an initial-state ensemble. The usual choice for the ensemble has been the full set of 2^N net states, each state weighted equally. Such a choice of ensemble is not indefensible, but it does have consequences that do not appear to have been fully recognized.

Consider the behavior of any of our switching nets. The 2^N-net states are naturally partitioned into separate groups, each group of states associated with a different cycle. Let each of these groups be called a confluent. In algebraic terms, the confluent in which a state is found is the transitive closure of that state. Each confluent consists of a (single) cycle, and all the states of run-ins to that cycle.

Since the probability of initiating a trajectory in a given confluent clearly depends on the total number of states in the confluent, if initial states are selected equiprobably and independently, all observed behavioral characteristics, such as frequencies of cycle lengths, also depend on confluent size. That is, using the typical initial-state ensemble, we do not select cycles or cycle lengths equiprobably; we tend rather to observe the cycles of larger confluents. We therefore derive the frequencies of cycle lengths weighted by confluent size. Note also that even when examining local cycle space topography, as for example, in examining the stability of cycles, since the cycles examined are generally not selected directly but are obtained via initial states, even here we examine the local topography of the more easily discovered cycles.

If the modeling context makes it appropriate, then the use of this particular, "flat," initial-state ensemble (equiprobable, independent choice of net states) is desirable. Such contexts are, of course, those in which it is sensible to think of the real system modeled as typically started at a fully arbitrary point in its behavioral repertory. We discuss this point at greater length in Chapter 6. It suffices here to point out that often enough the modeling context does *not* call for this kind of initial-state ensemble. To bring behavioral data generated by the use of the flat initial-state ensemble to bear on modeling contexts that do not directly fit its use naturally requires that allowance be made for confluent size effects.

From another point of view, however, the use of the flat initial-state ensemble appears quite reasonable. This point of view has apparently been adopted in many of the existing empirical studies on nets, in which the use of this ensemble is justified, often only implicitly, as a "first-look" effort at describing net behavior. That is, since little is known about such nets, almost any clearly defined method of eliciting the facts provides a gain in perspective regarding them.

5.1.3 Determining Cycle and Run-In Lengths

This section discusses technical problems in determining run-in and cycle lengths empirically. Readers not interested in details of search techniques should skip to

Sec. 5.2.

We always assume that the basic datum used in the determination of a run-in and a cycle length is a behavioral trajectory--a sequence of computed net states starting with some chosen initial state and ending with the last state simulated. Two general problems are: How many states should be calculated, and how should one go about determining whether a cycle exists in the trajectory? The two problems are related. Basically, we want to specify a relatively efficient procedure for producing and disclosing a cycle. In searching for a cycle, we will consider only methods that directly compare states.

There are two techniques to be considered here. Both make heavy use of the last computed net state, since a repetition of the last state in the trajectory is necessary and sufficient for the disclosure of a cycle. If a cycle exists (in the computed trajectory), the last state will be repeated; if the last state is repeated, a cycle is disclosable.

The first technique, "immediate search," is more straightforward. It takes each newly computed net state and searches the previously computed trajectory for a repetition of that state. As will be seen, this procedure minimized the number of states that must be calculated in order to disclose a cycle but may squander computer time spent in searching. The second search technique, "delayed search," calculates a preset number of states and only then searches for the cycle. This method is better at the task of finding the cycle, but can require calculating an excessive number of net states. The production method we have used is a mix of the two techniques.

We will compare how these two procedures generate information sufficient to determine run-in and cycle length. We will largely concentrate on the number of net states calculated and the number of net state comparisons made by the procedures. To avoid excessive detail, not all contingencies will be considered, nor will we attempt to specify absolutely optimal procedures.

As a first comparative datum, note that the calculation of a net state and the comparison of two net states for equality in general require very different amounts of computing resource. As a rough comparison, an early program written in assembly language for the IBM 7094 executed approximately 5800 instructions in calculating a net state (for a net of 75 elements), compared with about 15 instructions in testing for equality of two net states. These numbers are only rough guides as to what will be required in practice, particularly in the case of the test-for-equality operation, since two net states are unequal if *any* pair of corresponding element states differs. The comparison procedure can be terminated on the first encounter with dissimilar elements.

We observe that there are at least two ways of using the last state in searching for a repeated state: one can compare the last state with others starting at the first state proceeding forward, or start at the next-to-last state and move backward along the trajectory. This now gives us four procedures to consider: immediate search, forward and backward; and delayed search, forward and backward.

Immediate Search. It follows for immediate search that since the trajectory is calculated state by state and the searching for a duplicated last state occurs immediately after each calculation, immediate search procedures disclose a cycle as soon as sufficient information exists, that is, when the trajectory length in states is the disclosure length, d, plus one. (Recall that the disclosure length is the sum of the run-in and cycle lengths.) This occurs when d states have been calculated, since the initial state ordinarily has been specified by means other than a net state calculation. It is important to notice that immediate search will have repeated its scan of each entire existing trajectory for each of the d - 1 states produced in building up a trajectory d states in length. If each search requires i state comparisons of the last state against all other states, the total number of state comparisons required, before the d + 1 (last) state is calculated, is

$$\sum_{i=1}^{d-1} i = d(d-1)/2 = d^2/2 - d/2$$

The forward and backward methods will therefore differ only in how many net state comparisons are made after the last state calculation. A simple diagram of a trajectory d+l states long will illustrate that the last search carried out forward from the initial state will require the run-in length, r, plus one, comparisons of the last state. The backward search will require exactly one cycle length, c, comparisons, as can be seen below.

Net states:	o	o	o	o	o	/	0	o	o	o	/	0
Number:	1	2	...		r		r+1	...	r+c		d+1
			Run-in			/		Cycle		/Last state	

In summary, the requirements for immediate search are as follows:

	Computations	Comparisons
Forward	d	$d^2/2 - d/2 + r + 1$
Backward	d	$d^2/2 - d/2 + c$

It is clear that in simulating a set of trajectories which show the same disclosure length, forward and backward searches differ roughly to the extent that the sum of run-in and the sum of cycle lengths differ in the set of trajectories. The difference is therefore related to the difference between the mean of run-in and cycle lengths. Where this is unknown, the choice between the two methods of search is indeterminate.

The time required for a given simulation run is roughly the sum of the number of net state computations and net state comparisons required, each weighted by their respective times. It can be seen that regardless of the comparative individual times, if disclosure lengths become long enough, there must exist some d* beyond which point most computer time will be taken up in comparison operations, owing to the d^2 term in the net state comparison formula. Beyond this point, time requirements are dominated by the d^2 term. As an example, using the IBM 7094 program mentioned above, for N = 75 the crossover point occurred at about d = 1000, and to increase the aperture from 5180 states to 6125 states more than doubled the time required to calculate and search a trajectory. Where long disclosures are anticipated, it therefore becomes of interest to examine procedures such as fully delayed search, which avoid a heavy dependence of running time on disclosure length.

Fully Delayed Search. Fully delayed search procedures have in common the calculation of the trajectory out to the full size of the aperture before any searching for repetitions of the last state is done. Thus they make

$a - 1$ net state calculations, and differ only in how the single search scan affects the number of net state comparisons. It is clear that to disclose a cycle, $d + 1 \leq a$.

To consider the number of comparisons in fully delayed searches, note that we can display the trajectory with a disclosable cycle in a natural sequence consisting of r run-in states, m complete cycles of c states, and a sequence of e "excess" cyclic states beyond the last full cycle, where $r \geq 0$, $c \geq 1$, $m \geq 1$, and $c > e \geq 0$. Diagrammatically, this looks as follows:

Net states:	o	o	o	o	o /	o	0	o /..../	o	0	o /	o	0
Number:	1	2			r	r+1		r+c			r+mc		r+mc+e
			Run-in			/ Cycle 1/			/ Cycle m /			"Excess"	

Fully Delayed Search, Forward. For both forward and backward fully delayed searches, we will find that the number of states compared depends (but only slightly) on whether $e = 0$ or $e > 0$. The procedure we consider here takes the last state and makes comparisons forward from the initial state to find the first appearance of the last state ($r + c$ comparisons, if $e = 0$; $r + e$, if $e > 0$), then makes c comparisons forward until the second appearance of the last state is found. At this point all cyclic states are identified, and the cycle length can be computed. The next step compares state pairs that are a distance c apart backwards from the second appearance of the last state until a mismatch is found (e comparisons, $e > 0$) or until at most $c - 1$ comparisons have been made ($e = 0$). This identifies the last run-in state. Thus, for the forward search, the number of net state comparisons made is

$$\begin{array}{ll} r + 3c - 1 & e = 0 \\ r + c + 2e & e > 0 \end{array}$$

At the stage where the first appearance of the last state has been found, an alternative procedure for determining the cycle length would be to look at states that are located at points which evenly divide the span between the last state and its first appearance in the trajectory. Where cycles are quite long, this procedure might be more economical. For example, if the span between the two were 103, since 103 is prime, only one comparison is necessary: the last state against the next-to-last state. If these two states were equal, the cycle length would be 1; if unequal, the cycle length would be 103. Whether this procedure is advantageous depends on how much computer time is required to determine divisors, and so on, as contrasted with the time to make c net state comparisons.

Fully Delayed Search, Backward. The procedure takes the last state and makes comparisons backward along the trajectory until the next repetition of the last state is found. This requires c comparisons, identifies all cyclic states, and allows the calculation of the cycle length. The next step is to continue comparing the last state, backward, but leapfrogging by a distance equal to the cycle length, until the first nonequality is reached. Where m is the number of complete cycles in the trajectory, this step takes m - 1 comparisons ($e = 0$), or m comparisons ($e > 0$). The first appearance of the last state has now been found. The next step is to pairwise compare the states that precede the first appearance of the last state with states that precede the last state by an equal amount, up to the point of the first nonequality ($e > 0$), or until at most $c - 1$ comparisons have been made ($e = 0$). This step identifies the last run-in state, and requires $c - 1$ ($e = 0$) or e ($e > 0$) state comparisons.

In summary, fully delayed searches require the following number of state comparisons:

	$e = 0$	$e > 0$
Forward	$r + 3c - 1$	$r + c + 2e$
Backward	$2c + m - 2$	$c + m + e$

Note that the disclosure length does not appear as a quadratic in either of the expressions for fully delayed search leading to its preference over immediate search. Since the difference between the number of comparisons for forward and backward search are hard to predict in advance, but may favor backward search, one would want to opt for backward search in production runs making use of delayed search.

Immediate Versus Fully Delayed Search. Looking only at searches carried out backward, in summary, we have the following numbers of net state computations and net state comparisons (using the typical case, $e > 0$, for fully delayed search):

	Computations	Comparisons
Immediate	d	$d^2/2 - d/2 + c$
Fully delayed	$a - 1$	$c + m + e$

Immediate and fully delayed searches differ in both state computations and state comparisons, and can differ markedly on both. To disclose a cycle, both searches require one to use an aperture $\geq d + 1$ on running a single trajectory. In the more typical situation, one runs a number, say p, of trajectories. Here the aperture used (usually for all runs) must of course exceed d + 1 for all of at least a substantial fraction of the trajectories to avoid an unacceptable waste of computer time. However, fully delayed searches will compute $p(a - 1)$ states while immediate searches will compute only $\sum_{i=1}^{p} d_i$ where $d_i < a - 1$. Nevertheless, as already mentioned, immediate searches will be subject to a rapidly growing number of state comparisons if d becomes large, in contrast to fully delayed search, which is sensitive mainly to cycle length. This quantity may increase with disclosure length, but naturally must be less than d, and in fully delayed search, in any case, it does not appear as a quadratic term.

This suggests that a mixture of immediate and fully delayed search might be useful: that is, a procedure in which search is undertaken more often than once at the end of the computation phase, but not as often as immediate search would require. Our production technique uses two intermediate searches for repeated last states, first when the l_1-th state is the last state computed. If no cycle is disclosed at this point, states are computed to l_2, where another search is conducted with the new last state, and similarly to the a-th state, if necessary.

The point of this mixed technique is to gain some of the advantages of both delayed and immediate search. Its use raises the question of how l_1 and l_2 should be determined. If D(x) gives the number of trajectories with disclosure length x, the number of states that are computed is

$$C(l_1, l_2) = l_1 \sum_{x=2}^{l_2-1} D(x) + l_2 \sum_{x=l_1}^{l_2-1} D(x) + a \sum_{x=l_2}^{\infty} D(x)$$

and the problem is to minimize $C(l_1,l_2)$ with respect to l_1 and l_2. If D(x) is a constant, $C(l_1,l_2)$ is easily shown to be minimized at

$$l_1 = a/3 + 4/3 \qquad l_2 = 2a/3 + 2/3$$

In practice, D(x) is unknown. To get some heuristic guidance on the placement of l_1 and l_2, we empirically examined the (approximate) minimization of several different D(x) forms. We assume that $D(x) = 0,\ x < 2$ and in accord with the sensible practice of setting a so as to make cycle nondisclosure unlikely, $D(x) = 0,\ x > a$. With all $a = 100$, the results were as follows:

D(x)	l_1	l_2
Truncated normal: $40 \exp(-0.002(x-25)^2$	30	52
Centered normal: $36.8 \exp(-0.00144(x-50)^2$	50	75
Flat	33	67
Triangular: $50 - x/2$	30	60
Ramp: $x/2$	50	75
Truncated Exponential: $25 \exp(-0.02x)$	20	45
U-shaped: $50(x-50)^2/2500$	20	40

Thus, for "sensible" distributions of disclosure lengths, those that bulk at the center or to the left of center of the aperture, a, l_1 can be set somewhere between $a/5$ and $a/2$, with l_2 set to split the remaining interval. Distributions that bulk to the right of center and presumably are heavily truncated at a indicate a situation where a should be increased.

5.2 Functionally Homogeneous Nets (Ashby Nets)

This section provides a brief review of findings on the behavior of ensembles of functionally homogeneous nets, which we call Ashby nets in honor of W. R. Ashby. We will draw heavily on Walker and Ashby (1966). As the discussion to this point has assumed, the nets examined are autonomous, clocked, binary-state machines. In this section, all elements in any given net are functionally identical; that is, they compute the same function, whatever it might be. The particular nets considered in this section have two inputs which are unrestricted as to their point of connection within the net. The inputs of any one element might, for example, connect to the output of that same element. All inputs are connected to some element's output. Additionally, the elements considered here have one other input which is always connected to the output of the same element. This input line will be referred to as the "feedback input." These nets have connectance k = 2 as far as structural modeling is concerned on an element-to-element basis, but are k = 3 from the standpoint of functional form. As with all nets considered in this book, the state transmitted is the state computed, delayed by one clock-time unit. The state of the element is defined to be the state of the delay at time t (i.e., the element's output).

The elements of the nets considered in this section can be diagrammed as follows:

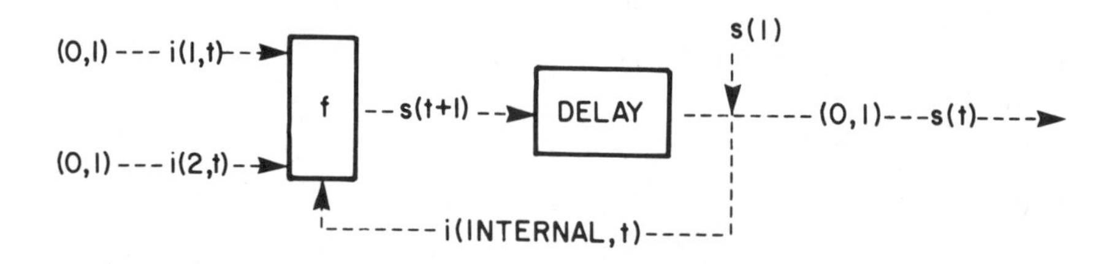

In the diagram i(1,t) and i(2,t) are states (either 0 or 1) carried by input 1 and input 2, respectively, at net time t; s(t + 1) is the state computed by the mapping f. This state will be the element output at net time t + 1. The feedback loop carries the output state back as an input (the feedback input) denoted i(internal,t). This s(1) line is to remind us that the initial state must be set by some means external to the computational process.

The mapping use in a net can be set out in the following form:

		i(internal,t)	
i(1,t)	i(2,t)	0	1
0	0	a	e
0	1	b	f
1	0	c	g
1	1	d	h

where a, b, ..., h are the element's output state at time t + 1, and are each 0 or 1 in any specific mapping. The lexicographic order used here appears in Walker and Ashby, wherein the inputs 1 and 2 are called the "left" and "right" inputs, respectively. They would be interchanged in the notation of Chapter 3.

Since there are three binary inputs to the mapping f, and the output is also binary, the number of distinct mappings is 256. Therefore, there are 256 different kinds of nets that can be considered, each associated with a different mapping. Using the by now familiar flat ensembles of structures and initial states to produce the population of behaviors descriptive of each net type, as mentioned above it can be shown that insofar as cycle and run-in length distributions are concerned, there are 88 distinct mappings that must be studied (Walker, 1971a; Walker and Aadryan, 1971). The ensemble of net states has 2^N distinct members, and the ensemble of structures has N^{2^N} distinct structural tables. Owing to the fact that the elements of the nets are identical, there are not in this case N^{2^N} distinct structural forms, necessarily. That is, the structural ensemble used in the Walker and Ashby study has some technical (i.e., nonmodeling) aspects to it. The "dilution" caused by the presence of similar structural forms is not severe, and as usual, the decision as to whether the ensemble used is appropriate for real-world modeling depends on the specific modeling context in use.

It is worth noting that one can remove the feedback loop in the element by a suitable choice of a mapping. There are, of course, exactly 16 mappings without a functional feedback loop.

5.2.1 Certain Measures of Ashby Net Elements

The point of the Walker and Ashby study was to survey the behavior of a class of complex nets made up from relatively simple parts. Since the study examined functionally homogeneous nets, attention naturally fell on the mappings used and on how mapping characteristics might be related to behavior. They defined four mapping characteristics: (1) **internal homogeneity**, already described, which reflects the general tendency of a mapping to output the same value; (2) **fluency**, which reflects the ability of a mapping to signal with either input fixed in value; (3) **memory** (the extent to which the columns in the table differ), which reflects the capacity of the mapping to treat inputs differently depending on changes in feedback values; and (4) **hesitancy** (simply a count of the number of column entries which match the column head in the mapping), which reflects the tendency of the mapping's output to remain unchanged, regardless of input activity.

A little reflection on the measures above will show that "fluency" is just about the opposite of "forcibility," as described in Chapter 4. "Hesitancy" is strongly related to the forcibility of the mapping via the feedback loop (if forcibility is counted only when like states force like states). In the feedback loop, if like states do force like states, then, owing to the loop, one has a lock-in situation produced by a mechanism exactly like Kauffman's "forcing loop" (Kauffman, 1970 or 1974), discussed in Sec. 5.3. Intuitively, all these measures are related to the tendency of the mapping to change or not to change state. Each mapping that is fixed in its output reduces the number of net states that can be explored by half; hence one would expect hesitancy, fluency, and internal homogeneity to have a significant effect on the length of cycles. It is not intuitively clear how "memory" (as defined above) ought to affect behavior, since even if the mapping columns differ sharply, the actual appearance of this difference in behavior will depend on whether the element actually tends to change its output value or not. Thus the effectiveness of "memory" should be related to levels of hesitancy. Indeed, the Walker and Ashby study did not find memory to be a very important variable on its own.

The behavioral data cited in this section are taken from Walker and Ashby (1966) unless otherwise noted. All nets examined in that study had 100 elements. Five different structures and 10 different initial states were used, all selected from flat ensembles. The aperture size for most of the runs made was 500 states. If no cycles were disclosed in any particular net, one trajectory was run with the aperture set to 5,180 states.

5.2.2 Disclosure Lengths

Recalling the earlier discussion of the size of reasonable apertures in comparison with the size of the behavior space of N = 100 nets, perhaps the first question is: On random jumps into the cycle space, is an aperture of 500 sufficiently large to disclose any cycles? That is, are cycle lengths typically small enough and are cycles located "close" enough to the transient states in the confluents of these nets, that a 500 state aperture will very often "cover" both the run-in and cycle? The answer is yes; in about 75% of the mappings, cycles are detectable over half the time. These mappings then, as a class, in homogeneous nets show a very sharp localization of behavior. There are very sharp differences between mappings with respect to disclosure length, however. Those mappings which did *not* often exhibit disclosure lengths of this restricted size typically show some resemblance to the logical functions "exclusive or" or "equivalence." Both of these functions have the characteristic that holding one input constant, it is still possible to transmit changes through the element via the other input. These mappings are the high-fluency mappings. From later data it can be seen that their disclosure lengths are long largely because of very long cycle lengths (Walker, 1976). Interestingly, their run-in lengths typically are very short.

A subsequent study looked in more detail at disclosure lengths (Walker and Aadryan, 1971). This study found that a measure similar to fluency which reflects mappings' sensitivity to changes at element inputs is most effective in predicting average disclosure lengths for given mappings. Hesitancy is second in importance. The quantity actually predicted well by these measures is the reciprocal of the median disclosure length. This quantity can be thought of as the average proportion of the total calculation required by the net to disclose a cycle.

5.2.3 Run-In Lengths

How long does it take to get to the cycle from random points in the behavior space? As with disclosure lengths, run-in lengths vary widely over the class of mappings. The smallest and largest run-ins observed here were zero and 5085 states. The typical distribution of run-in lengths for those mappings with a fairly high number of disclosures was found to be skewed toward longer run-ins. All the nontrivial mappings showed relatively few run-ins of length zero, many short run-ins, and a scattering of relatively long ones. From this we can say, at least for those mapping showing relatively short disclosures, that their larger confluents have more transient than cyclic states, and that most transient states are located rather close to the cycle, with fewer transient states farther from the cycle. As to what general influences affect run-in lengths, maps with higher internal homogeneity tended to show shorter run-ins. Both extremes of hesitancy tended to produce shorter run-ins.

More recent data (Walker, unpublished data) show that the mappings with the very longest disclosures tend to have very short run-ins, with a substantial proportion of length zero run-ins--30% or so. Although the net sizes used to obtain the data were not large (N = 17 was the largest), the finding was relatively consistent for net sizes up to that size of net. Therefore, these particular mappings tend to have confluents in which relatively many states are cyclic states. The remainder of the states in the confluent are grouped very tightly around the cycle.

5.2.4 Cycle Lengths

For the observed class of N = 100 Ashby nets, cycle lengths disclosed ranged from 1 to 4040 states. The shortest cycle length possible, the length 1 state, was observed fairly often. Taking into account the trajectories in which no cycle was disclosed, and which may end in cycles of length 1, the proportion of trajectories ending in the shortest cycle was estimated to lie between 1/3 and 1/2. It is known from later data that 1/3 is the more descriptive of this class of nets because the long disclosure mappings tend to have relatively long cycles.

The mappings used are also an important source of variability in cycle lengths. About 20% of the 256 mappings exhibited almost no cycles other than the shortest. Some other interesting regularities were found among the cycle lengths of given mappings. For example, some mappings showed cycles whose lengths, while differing, were all multiples of a common factor [greatest common divisor (GCD)]. The GCDs observed were 2, 3, 4, and 8. These GCDs are not exclusively a matter of the particular net size used, showing some constancy across changes in net size (Walker, 1971a). A few mappings showed a tendency to produce long prime cycle lengths which were sensitive to structural change in the nets.

What aspects of mappings contribute to types of cyclic behavior? Maps with higher internal homogeneity are more likely to show smaller cycles. Similarly, mapping with high hesitancy (forcible on the feedback loop with like states forcing like states) also are likely to show small cycles. Mappings with intermediate hesitancy tend to show longer cycles. As mentioned above, long cycles are also produced by mappings which are sensitive to changes in input values (i.e., high-fluency mappings). The longest cycle we have observed is 27,559 states in length, in a N = 17 net using a high-fluency, high-memory mapping. This particular mapping is the binary adder (of all three inputs). The longest cycle lengths are on average produced by the mapping which is the negation of the binary adder just described. Based on its behavior as N is increased, it can be said that fairly frequently this mapping produces a behavior space which has only two cycles: one of length 2 and one of length $2^N - 2$--a behavior space which has in it only cyclic states.

5.2.5 Behavior as Affected by Connectance

Although the nominal connectance of this class of Ashby nets is fixed at k = 3 (with one feedback input), as remarked earlier, it is possible to modify effective connectance by functional manipulation. Here that manipulation is simply a matter of choosing the mappings of the net. By choosing correctly, one can obtain any (fixed) connectance up to 3. Given the purpose of this book, the mappings of greatest interest are those with no feedback inputs. Of the 256 mappings in this k = 3 class, 16 have no effective feedback input. Of these, 2 have no effective input at all (the trivial maps), 4 have exactly one effective input, and 10 have exactly two effective inputs. (As a matter of interest, a forcible input is an effective input, but an effective input does not necessarily force, in the case of the nontrivial mappings.) The means of run-in and cycle lengths of these no-feedback classes for N = 17 are as follows (data from Walker, 1976):

Connectance	Run-in length	Cycle length	Number of mappings
0	1.0	1.0	2
1	4.2	4.2	4
2	4.4	162.9	10

Neglecting the two trivial maps, the mean run-in lengths differ little between the effective connectance classes tabled, while the cycle length mean for the connectance two class is markedly larger. This effect results from the very large cycle lengths shown by 2 of the 10 mappings in the class.

An effective connectance of 3, in this class of mappings, at N = 17, produces a run-in length mean of about 9, and a cycle length mean of about 15. But it should be remembered that an effective connectance of 3, for these elements, requires an effective feedback input. Of the 256 mappings in the whole class, the 240 with an effective feedback input can be separated into 60 which force via the feedback input. In that 60, like values force like values in 31. As mentioned earlier, it would be expected that these 31 mappings would have short cycles. They do. None shows any but cycles of length 1, and the same is true at N = 100. Thus these mappings at k = 3, as a group, having 12% of mappings in which like values force like values on a feedback input, may not resemble classes of k = 3 mappings which do not have feedback inputs.

5.2.6 Topography of the Cycle Space

Are switching nets' behavior spaces in some sense smooth and continuous surfaces on which it makes sense to define distances? Or are they merely collections of intersecting trajectories which have in common the cycles they share? This is an important point as regards modeling. The interpretive scope of switching nets clearly appears larger in the former case as against the latter. We discuss this consideration in more detail in Chapter 6, using the modeling issue here to motivate our examination of the data.

Intuition is not very informative in answering the question posed. One might expect a certain combinatorial specificity of behavior resulting, perhaps, from the discontinuous nature of the behavior space, or, depending on the mapping, from nonlinear aspects of elements' functioning. On the other hand, one might also suspect that if, for example, a mapping is biased toward a certain output value, a behavioral bias toward similar outcomes at the whole-net level might be produced.

Distributions of cycle and run-in lengths, by themselves, only provide hints as to cycle space topography. Some of these have been remarked on above, but hints they remain. What do the data say? Earlier studies in complex systems, such as those by Ashby (1960) and Fitzhugh (1963), as well as Walker and Ashby, found that the "distance" between successive trajectory states, in nets typically terminating in cycles of length 1, more or less progressively decreases to zero. The distance referred to is, as in Sec. 4.7, the Hamming distance (the Minkowski metric with index 1, or the "city block distance"), here the number of net elements that change value in moving from one net state to the next. This result, although still describing single trajectories, because of its frequency of appearance, does suggest that there may exist what might be called stability neighborhoods around the cycles. The term "stability" is appropriate. The suggestion is that if such a net were cycling, a relatively small displacement (as by an error in computation in one element) would very likely be followed by the net's return to the cycle, since there appear to be so many trajectories in which states close to the cycle in order of succession are also close to the cycle in the sense of the distance metric (recall the stability discussion in Sec. 3.3). If the displacement were larger, the suggestion is that the likelihood of the net's return to the cycle would be less.

Kauffman (1969) explicitly examined this kind of stability, but in functionally heterogeneous nets. His results are reviewed in Sec. 5.3. Kauffman's study was followed by one that examined the stability of cycles in Ashby nets using no-feedback elements (Walker, 1971b). This study found that most of the no-feedback mappings are highly stable, regardless of the cycle length, under small displacements from the cycle. The stable mappings produced a return to the cycle 80% of the time or more, under unit displacement in N = 100 nets. The two high-fluency mappings showed about 30% of returns to the cycle. (This 30% figure is for cycles disclosable at N = 40 with an aperture of only 100 states. The result is a set of cycles biased toward smaller, and hence probably more stable, cycles.)

We mentioned above that the latter mappings, from run-in length data, apparently have behavior spaces in which run-in states are tightly grouped (in the "which state follows which state" sense) around cycles, at perhaps two run-in states per cyclic state. Since there are N states lying 1 unit distance from any net state, of which only about two are on run-ins to the given cycle, the observed stability result for the high-fluency mappings is not surprising. Moreover, given that the ratio of roughly two run-in states to each cyclic state does not appear to change much as net size increases, one could also expect that the likelihood of a displaced trajectory returning to the cycle would decrease with increasing net size. This appears to be the case. For the other no-feedback mappings, larger nets provide greater stability rather than less.

By and large, the no-feedback mappings show decreasing stability with increasing displacement. This further indicates a topographical regularity. States farther from the cycle are less likely to be associated with that cycle.

More recent work examining stability of mappings with feedback has been done but is at this time not analyzed fully. However, the general impression provided by the data is that the mappings with feedback show less stability. Among these mappings some also show what might be called paradoxical topography, in the sense that they show stability that may be low (or declining) for small displacements, but with still larger displacements, they show an improvement in stability. This indicates a fairly complex topography at medium distances from the cycle.

Another interesting question of cycle space topography as examined through cycle stability has to do with where in the cycle the displaced trajectory regains the cycle. If a mapping is highly stable (under displacements of a given size) does the point of return to the cycle show regularity? The question of where the trajectory returns can be considered with respect to the point at which the original trajectory entered the cycle, the point at which the displacement occurred, or the point at which the net would have been when the return occurs. We have looked at these questions in homogeneous nets only enough to discover some hints of regularity, and plenty of variability. The variability is sufficient to make it clear that in looking at these phase relations one should give more attention to single cycles, in the sense of more exhaustively dealing with each cyclic state, rather than choosing one state at random and displacing the net from that state. We have more to say on phase of return in Sec. 5.3.

Cycle Length Versus Run-In Length. Another datum that can be considered to give some picture of cycle space topography is the correlation between cycle and run-in length. That is, if larger cycles generally show longer run-ins, then on average, trajectory arborization is greater farther from the cycle in proportion to the cycle length. As it happens (Walker, 1973), most (87%) of our 256 mappings show correlations between cycle and run-in length that are low. The remaining show correlation magnitudes that are less than 0.51. Since there appeared to be no clear nonlinear relations between run-in and cycle length, more generally it was found that trajectory arborization is not very systematically related to cycle length.

Looking only at no-feedback mappings, 10 show low correlations, and four show moderate negative correlation between cycle and run-in length. Among the latter four are the two high-fluency, no-feedback mappings. These mappings show a moderate tendency toward shorter run-ins with longer cycles. This is not surprising in view of the tendency of such mappings to exhibit cycles of such size that few run-in states can exist in the behavior space.

Density Data. Still another way of characterizing location in the behavior space is by using the number of 1's in the net state. This is a metric in the sense that it measures the Hamming distance of the given state from the "all zeros" state. Since, using this scheme, one is referring all measurements to the same point in the space, one can conceive of the measure as an indication of location. To generalize the measure to behavior spaces of different sized nets, the number of 1's is usually divided by N, yielding "density" (of 1's). Walker and Ashby used a simple probability model to predict terminal densities, that is, the densities expected in cyclic states, in the present class of nets. The simple probability model they used assumed, contrary to fact, that at any given time the net state density of 1's can be used to assign probabilities of input configurations. These, using the mapping itself, provide probabilities of output 1's, and hence subsequent densities. In general, the relationship between density at time t and density at time t + 1 using the model is a third-degree polynomial in density at t. Invoking the assumption of independent probabilities, which is equivalent to saying that the net structure is randomized again after each time interval, the probability model can be used to predict the terminal densities of the mapping. With regard to the present class of mappings, Walker and Ashby only remark that for some mappings terminal densities are close to those predicted, whereas for others the discrepancies are large. (Walker and Ashby were interested primarily in the cycle and run-in characteristics of the mappings.) One intriguing question raised by this observation is: What aspect of mapping is it that allows statistical behavior to be manifest in nonstatistical (fixed structure) models? Reviewing density data for the no-feedback mappings suggests that higher internal homogeneity may be associated with increased statistical predictability. Unfortunately, it is also true that internal homogeneity is not the only factor effective here. Walker and Ashby give an example of a mapping with low internal homogeneity for which net state densities in cyclic states are "close" to the statistical prediction. In any case, it is clear that "where" cycles are in the behavior space can be of interest in some modeling contexts. We return to this matter in the next section.

5.2.7 Effects of Net Size

The introduction of this chapter calls attention to the sharp effect of net size on the size of the behavior space, and to the difficulties that impede empirical examination of large behavioral patterns in the behavior space. Net size, then, is both theoretical and practically interesting. Does net size have predictable effects on behavior? Can small, and hence more easily examined nets' behavior be used to predict how larger versions of the net will behave?

One study (Walker, 1971a) looked at the effect of net size change on the greatest common divisors of cycle lengths. Using data from the Walker and Ashby study, where N = 100, a similar design with N = 75 produced data for comparison. The N = 75 nets necessarily used different net structures and initial states. Under the sampling conditions used (5 net structures and 10 initial states for each mapping at each net size), the results were that the GCDs of the cycle lengths of both net sizes matched for 70% of the mappings. For the net sizes used, there appears to exist some constancy of cycle lengths' GCD. Perhaps a better way of stating this result is to say that we have some indication that the spectrum of cycle lengths produced by a mapping appears to be relatively constant across modest changes in net size.

A speculative approach to the problem of predicting behavior at greater net sizes using smaller nets was stated by Walker (1971a). Briefly, the idea was that "size" may be essentially a structural characteristic. Hence, if some aspect of behavior is affected mainly by structural factors, that behavior should be related to size as well. Considered explicitly in Walker (1978), the conjecture failed for all variables tested: cycle length, run-in length, and disclosure length. The total variance of a variable at smaller net sizes did as well in predicting size effects as did the variance of the initial state or the variance due to structure. In fact, the most powerful and perhaps the most direct way of predicting how well a mapping might respond to change in net size, in comparison with other mappings, proved to be by using relative comparisons of means at the smaller size as predictors of relative sizes of means for the larger nets.

A more recent study of cycle length distribution (Gelfand and Walker, 1977) puts it this way:

> The 88 transformations can be more or less classified into two groups; those which are essentially degenerate over N and those which exhibit changing, often rather volatile, behavior over N. The former group is largely made up of transformations which exhibit almost trivial behavior in the sense of yielding cycle lengths which are virtually always one, always two, or a relatively stable proportion of one's and two's regardless of N. The latter group contained transformations which clearly seemed to be varying in N. . . .

From this account and those preceding, we can summarize for cycle lengths, as follows. Variability appears to be a key to understanding at least the grosser aspects of cycle length behavior over N. Those mappings which show relatively greater variability at small net sizes tend to show greater variability at larger net sizes. Since all of the mappings exhibit cycle length distributions that extend down to very low values, at all net sizes observed, cycle length mean tends to be correlated positively with variability. That is, for most of the class of mappings, variability and means are not independent. Mappings for which cycle length variability and means are independent over N are those mappings for which variability is zero, or very low. Hence it is not surprising that means at small net sizes can be used to predict means at larger net sizes. Although not analyzed at this time, a similar relationship appears to hold for run-in lengths. (Run-in length is not necessarily related to cycle length, over the class of mappings, in any obvious sense. For example, the mapping with the largest mean cycle length is not the mapping with the largest mean run-in length. In fact, the opposite is more nearly the case, as the reader may recall from an earlier discussion.)

5.3 Functionally Heterogeneous Nets (Kauffman Nets)

This section reviews empirical studies of ensembles of functionally heterogeneous nets. The nets discussed are, as before, clocked, autonomous, binary-state machines with irregular but fixed structure. The nominal connectance of all the elements in a net is the same. The mappings used in a given net, however, can and usually do differ from one another, hence the term "functionally heterogeneous." The empirically observable behavior of ensembles of these nets has been studied extensively by Kauffman, most often in a biological context. Kauffman's behavioral findings are reviewed below. His interpretive contributions are discussed in Chapter 6.

The elements of the Kauffman nets considered in this section are exactly like the Ashby net elements diagrammed in Sec. 5.2 if one removes the feedback loop. It is clear that this need not necessarily be done. That is, a functionally heterogeneous net *could* make use of feedback elements. Those studied to date simply have not added this complication to the model. For that reason, the mappings of the Kauffman net elements have only two inputs, and are exactly equivalent to the 16 no-feedback Ashby net elements mentioned on occasion in Sec. 5.2.

The ensembles used in studying Kauffman net behavior include those of net structure and initial state already familiar from the discussion of Ashby nets. Since Kauffman nets are functionally heterogeneous, a new ensemble is added, namely an ensemble of functions. There are several functional ensembles of interest in this section. Understandably, in his earlier studies, Kauffman used the bench-mark functional ensemble of all 16 functions, equally weighted, a flat functional ensemble. We have called (Sec. 4.3) nets sampled equiprobably from flat structural, flat initial state, and flat functional ensembles completely or fully random nets. The reader should keep in mind that nets we describe as fully random are still structurally rigid, and that the flatness of the structure ensemble refers really to the tables of assignments of inputs to outputs, not to structural forms as such. Kauffman occasionally modifies the flat functional ensemble by deleting the two trivial mappings.

Functional constraints examined below include functional ensembles described by forcibility, extended threshold, and internal homogeneity. Our discussion of Kauffman nets proceeds by taking up, in turn, different types of functional ensembles.

5.3.1 Fully Random Kauffman Nets

As described earlier, a fully random Kauffman net is a net in which connections among elements and assignments of mappings to elements have been chosen equiprobably from flat structural and functional ensembles. Kauffman generally used the customary flat ensemble of net states in selecting the initial states of behavioral trajectories.

Behavior As Affected by Connectance. Kauffman has considered nets with inputs from all elements (a subset of nets with connectance $k = N$), nets with $k = 1$, nets with $k = 2$, and $k = 3$. He rules out from further consideration as plausible models both totally connected nets and $k = 1$ nets. He concludes that they have behaviors which are biologically impossible: very long, unstable cycles (Kauffman, 1969, 1970). He therefore concentrates on $k = 2$ and $k = 3$ nets.

In view of Sec. 4.3 we have no reason to question Kauffman's general characterization of totally connected nets (based on Rubin and Sitgreaves, 1954). It is worth pointing out that a characterization of the totally connected net (or any other net for that matter) as to its modeling relevance probably should take the median cycle length more seriously than the expected cycle length as a measure of relevance, when the evidence suggests a strong positive skew in the distribution. Nevertheless, for the totally connected class, with the expected cycle length an exponential function of N, even if the median is only a small fraction of the mean, the class distribution has typical cycle lengths large enough to throw doubt on the entire class as a source of empirical models for even moderate sized nets.

As to the $k = 1$ class, Kauffman (1969, p. 443) writes:

> State cycles arise whose lengths are a maximum of two times the lowest common multiple of the set of structural loop lengths. For random nets as small as 100, the state cycles generally exceed millions of states in length (Sloane, 1967). One-connected random nets

> possess behavior cycles capable of realization by no earthly organism.

This conclusion almost surely overstates the case.

Holland (1960) showed that in nets such as our k = 1 nets, state cycles arise which are *divisors* of twice the least common multiple (LCM) of the net's structural loop lengths. How often the maximum cycle length mentioned by Kauffman actually occurs is another matter entirely. On that point, Kauffman's reading of Sloane (1967) appears in error. Sloane's purpose was to propose nets that have long cycles. Indeed, his essay can be interpreted as pointing to how difficult it is to provide really long cycles in k = 1 nets. Furthermore, neither our Ashby net data for functionally one-connected nets nor our Kauffman net data support Kauffman's long-cycle conclusion for k = 1 nets. See, for example, Fig. 5.1, where original data for fully random Kauffman nets are plotted as a function of net size and connectance. Over the net sizes studied, it seems clear that connectance of one does *not* typically produce excessively long cycles.

On the other hand, Kauffman also proposes important *interpretive* reasons (particularly in his 1970 paper) for considering k = 1 nets relatively useless as biological models (see the discussion in Sec. 3.5).

Moving to fully random nets with k = 2 and k = 3, we can support Kauffman's finding that such nets are biologically reasonable as judged by behavior.

Cycle Lengths. According to Kauffman (1969), two connected nets (with the two trivial mappings disallowed) show a *median* cycle length (Kauffman's own analysis often made use of medians) that is roughly the square root of the net size. This is a vast reduction in length compared with what could be the case. Allowing

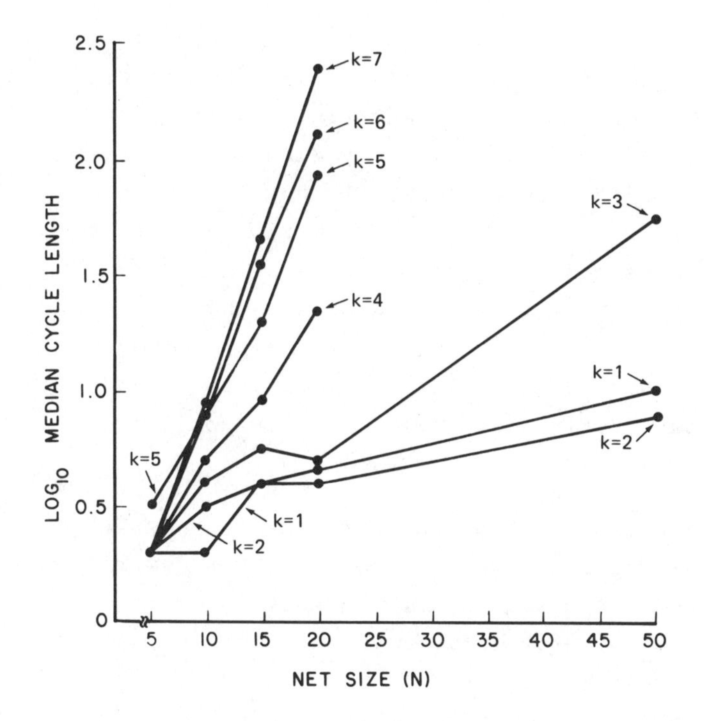

Fig. 5.1. Log median cycle length as a function of net size and connectance for fully random Kauffman nets. For each data point, 100 trajectories were initiated and at least 96 cycles disclosed. Trivial maps excluded. Original data.

the trivial mappings in the ensemble has the effect of further reducing the median. That is, the typically small cycle length of the k = 2 fully random net is not exclusively a contribution of the trivial mapping's presence. This finding is consistent with our Ashby net data. Sherlock (1979) argues that for such nets a roughly inverse relationship exists between cycle length and the number of cycles.

Kauffman's 1969 data indicated that k = 3 nets had slightly longer cycle lengths, approximately N, still very restricted in length.

Run-In Lengths. Kauffman found run-in lengths in k = 2 fully random nets to be uncorrelated with the length of the cycle to which the transient ran. Overall, Kauffman's run-in lengths (in k = 2, fully random nets) appear comparable to, but perhaps somewhat longer on average than his cycle lengths. He remarks that, as for cycle lengths, the distribution of run-ins is highly skewed, having a long tail toward greater lengths. (Kauffman describes the distribution as "highly skewed toward short lengths," but it is clear that he intends to describe what statisticians would call a *positive* skew.)

Cycle Space Topography. Kauffman has drawn a very complete picture of cycle space topography for the k = 2 fully random net. The regularities observed are remarkably broad in character, sufficiently so, in fact, to allow him to base much of his detailed case for the biological relevance of these nets on their behavior space topography.

For example, within the typical cycle, most net elements do not change their state, hence cyclic states are highly similar. At a distance of one state from the cyclic states lie states which have a high probability of being on run-ins to that same cycle. Thus cycles are highly stable under perturbations of 1 unit. The observed probability of return to the cycle under unit displacement is around 90%.

An empirical case can be made that k = 2 nets are especially stable (see Fig. 5.2). These original data give the percentage of returns to the cycle under unit displacement for fully random Kauffman nets of different size and connectance. Since percents are plotted, with the sample sizes used here the standard error of the individual point is roughly 5 percentage units. This

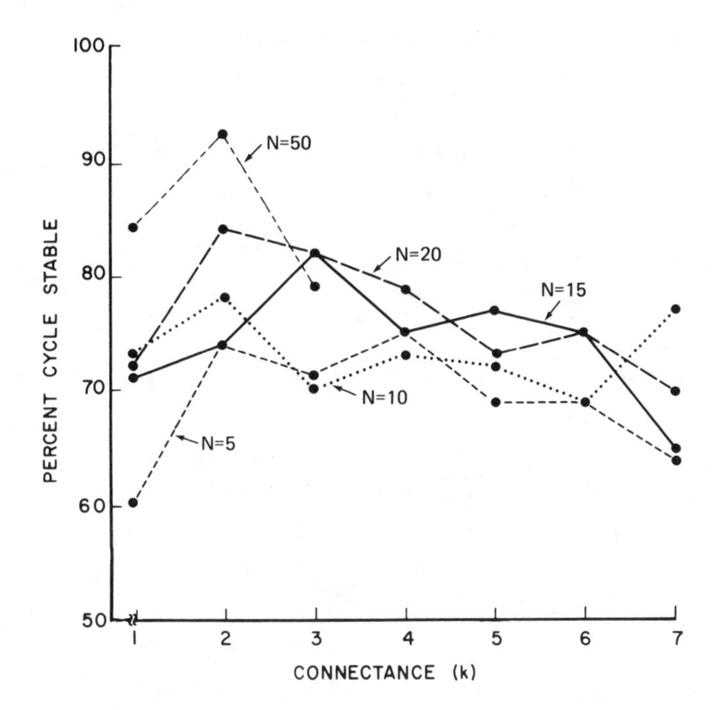

Fig. 5.2. Stability of cycles of fully random Kauffman nets as a function of connectance and net size. Unit displacement was used in all cases, one displaced trajectory being initiated from each originally disclosed cycle. Plotted points give the number of displaced trajectories that re-encountered the original cycle as a percent of the number of displaced trajectories initiated. See legend of Fig. 5.1 for sample sizes (number of original cycles disclosed). Trivial maps excluded. Original data.

makes it difficult to be confident of particular comparisons in the figure. However, stabilities increase for all five function lines from k = 1 to k = 2, and four out of five decrease from k = 2 to k = 3. Although we may be capitalizing on sampling error in remarking on these observations, additionally the envelope of the function lines suggests that a peak in stability exists around k = 2 as Kauffman (1970) suggested. Note, however, that k = 1 nets do not appear to be as unstable as Kauffman believed them to be. One can also see some suggestion of increasing stability with increasing net size. Perhaps the most surprising finding in these data is that stability is fairly high over the entire range of connectance values studied, k = 1 through 7. This is so even for the N = 5 case suggesting that the low stability Kauffman remarks on (1970) for k = N nets may hold only for large nets.

Returning to a question raised earlier: In cases where the cycle is stable, *where* does the displaced trajectory reencounter the cycle? Considering only the most stable fully random Kauffman net we have studied, k = 2, N = 50, we find that this net typically returns to the state of the cycle *at which it would have been had there been no displacement*.

To fix these ideas, take a particular example. In the set of nets under discussion, the first cycle disclosed had a length of 16 states on a 39-state run-in. The stability of this cycle was then tested in the following way. A state on the cycle was chosen at random--the eighth. A net state 1 unit distance from this latter state was chosen at random and let be the start of a new trajectory. In two steps from this displaced point, the net was back on the original cycle, at the state where it would have been, state 10, if the displacement had not been made. The reencounter point in this case is two states beyond the state chosen for displacement and seven states ahead of the cyclic state reached by the original run-in. Converting these phase descriptions to percentage of the cycle length, the reencounter occurred at -44% with respect to the first state in the cycle, +13% with respect to the state "displaced," and 0% with respect to where the net would have been. Now considering all the data, averaging the percent phase for the 99 cycles disclosed (but neglecting in this the eight observed cycles of length 1), we obtained (mean, standard deviation) (2.8, 26.5) for the first cyclic state, (16.8, 24.7) for the displaced state, and (0.8, 10.3) for where the net would have been. The standard deviation is more diagnostic in this set of values. If percent phase is spread evenly over all possible values, the standard deviation is around 30 and the mean is around zero. Hence reencounter phase with respect to where the net would have been is quite low on average. In fact, on 74 occasions (out of the 83 represented by the stable cycles of length more than 1) the net returned to exactly where it would have been. The k = 2 fully random Kauffman net not only returns to the cycle with high probability under unit displacement, but it typically returns in synchrony with net activity that existed earlier.

Although it would be advisable to study these behavior space properties in fuller detail, the suggestion is that low connectivity nets have behavior spaces in which, close to cycles, trajectories spiral into the cycle.

To return to Kauffman's data, after making allowance for the fact that this search for cycles makes use of a comparatively limited number of initial states, Kauffman estimates the median number of *distinct* cycles in a net's behavior space to be the square root of the net size. That is, the typical k = 2 fully random net has as few distinct cycles as its typical cycle lengths are short. The distribution of the number of cycles per net, for a fixed net size, is skewed in the positive sense. That is, there are many nets with rather small numbers of cycles, and progressively fewer nets with larger numbers of cycles.

How are cycles themselves grouped in the behavior

space? Measuring the distance between two cycles as the distance between the two closest states, Kauffman found that this minimum distance ranged between 1 (the smallest value possible) and 30% of N, with the median about 5% of N. He also found that when a net showed many cycles, they often clustered into groups of cycles showing very small intragroup clearances, and much larger clearances between groups.

These cycle groupings suggest (but do not ensure) that unit perturbations of the net state would shift the net from one cycle to only a few, rather than many, other cycles. This condition of "restricted local reachability" proves to be the case. The net, under unit perturbation, will shift from a given cycle to only one to six others with probabilities between 1 and 5%, and to a few other cycles with much lower probabilities. More surprising is that the transition probabilities are not symmetric; being relatively easy to move from cycle A to cycle B does not mean that it will be easy to get back to cycle A from cycle B.

Density Data. To get additional information on "where" in the behavior space fully random Kauffman nets operate, and how far apart net states on the trajectory are, we ran a study (Gelfand and Walker, 1982) which looked at net sizes of 12, 36, and 108, for connectivities of 2, 3, 5, and 7. "With replacement" sampling of inputs was used. This is the same scheme as that used by Kauffman to generate all his data. One hundred trajectories were started randomly at each of three initial density levels: 25%, 50%, and 75% 1's in the initial states. In that study, we recorded at each net time densities of 1's in the net state and densities of changes (i.e., the number of changes divided by N) in element values between that and the previous net state.

Means and standard deviations of the data of each of the 100 trajectory results for each starting density were calculated. The trajectories were run out to T = 25. We found that the densities of 1's in the initial states made no systematic difference in the results. The data for the twenty-fifth net state, in Table 5.1, have therefore been averaged over the three density conditions.

Since the model developed in Sec. 4.7 assumes inputs to be sampled without replacement, the theory of that section applies only approximately here (i.e., in the sense that when N is large and k is small, sampling without replacement is essentially equivalent to sampling with replacement). The k = 1 results of that

Table 5.1. Density of 1's in the 25th State, and Density of Element Changes (Normalized Distance) Between the 24th and 25th States

		Density of 1's, T = 25		Density of changes, T = 24 - 25	
N	k	Overall mean	SD mean	Overall mean	SD mean
12	2	0.499	0.144	0.182	0.184
	3	0.497	0.142	0.307	0.211
	5	0.507	0.140	0.419	0.184
	7	0.483	0.144	0.467	0.159
36	2	0.501	0.088	0.167	0.151
	3	0.500	0.083	0.347	0.138
	5	0.502	0.081	0.477	0.083
	7	0.495	0.082	0.492	0.083
108	2	0.494	0.048	0.113	0.097
	3	0.503	0.046	0.370	0.078
	5	0.496	0.047	0.485	0.047
	7	0.501	0.049	0.494	0.048

Overall means are averages of the means of the sets of 100 trajectories for 25%, 50%, and 75% initial densities. Similarly for the means of the standard deviations. Source: Original data.

theory, of course, directly apply here. See Sec. 5.3.4 for a discussion of results relevant to the theory.

The time course of density of 1's is easy to summarize. All the nets studied, regardless of initial density, immediately acquired and varied around mean values of approximately 0.5, with standard deviations that can be seen to be close to $\sqrt{N/2}$. That is, using with replacement sampling of inputs, fully random Kauffman nets, starting from initial states not too rich or deficient in 1's, on average operate at densities of 0.5, with variability a function of net size.

The observed time course of density of changes--the average number of changes per element from one trajectory state to the next--is more complicated. It decreases from an initial level to an apparent terminal level, following a curve similar to but below that shown by fully random Kauffman nets under without replacement sampling of inputs. The effect of with-replacement sampling, compared to without-replacement sampling, is largely to decrease distance between adjacent net states. See Sec. 5.3.4 for without-replacement data.

The terminal average distance between net states is a function of connectance and net size. High N, high k nets' distances between trajectory states appear to average 0.5N. For two-connected nets, the terminal average distance between net states appears to decrease with net size. For nets of size 108, two-connected, the terminal distance between states averages approximately 0.113N, with standard deviation around 0.048N. The latter values match closely Kauffman's 1969 findings. The correspondence is to be expected even though Kauffman's data are based on a different initial state ensemble, because of the observed insensitivity of our terminal levels to the starting states used.

5.3.2 Kauffman Nets and Forcibility

Kauffman, particularly in his 1970 and 1974 articles, examines the effect on behavior following from controlling the functional ensemble via forcibility. The term "forcibility" has been defined in Chapter 4 and follows Kauffman's usage exactly. To review briefly, mapping is forcible on a given input if one state on the input guarantees that the mapping output will assume a given value regardless of the states on other inputs. The "forced state" of a mapping is the state guaranteed. It is guaranteed by the presence of a forcing state on any of the forcing inputs on the given mapping.

Anticipating the behavioral effects of control by forcibility requires some attention to network structural details. Following Kauffman in what follows, a mapping A "forces" mapping B if and only if

1. A is forcible on one or more of its inputs.
2. B is forcible on the input line from A, by A's forced value.

Given the above and recalling that the forced value of any forcible element is the same regardless of the input that may be forcing at any given time, we can go on to consider what happens when a net contains chains of elements which are pairwise forcible as just defined. Such chains are "forcing structures" and may or may not contain loops.

Forcing structures are such that

1. A forcing state introduced to any element in the structure forces all descendent elements to their forced state, since the forcing process proceeds deterministically down the chain.
2. The forced state may differ at different points in the structure.

3. Any forcing loop is a positive feedback loop with a maximum of two steady states (cycles of length 1).

On the latter point, a stable steady state exists when each element is in its forced state. This condition clearly follows the introduction of a forcing state on the inputs of any element of the forcing loop. In this condition, the loop is insensitive to all external events. The loop's descendent forced elements themselves become fixed in their forced states. Some loops have an externally sensitive steady state in which all elements are in the states opposite to those of the stable steady state. The latter steady state can persist, and thus be part of a net cycle, only on net trajectories such that no member of the forcing loop ever changes state.

Clearly, forcing loops strongly constrain the behavior of a network. Forcing loops exhibit a strong tendency to return to the status where all elements are forced, if ever perturbed. Something like this homeostatic tendency also exists in loop-free forcing structures. As mentioned above, once a forcing value impinges on an element in the loop-free structure, forced states propagate down the structure. Moreover, forcing values can enter the structure ahead of the wave front initiated by the first forcing state. This creates a strong tendency for later members of the structure to be in their forced values all, or more of the time.

The extent of a forcing structure and its configuration, then, determine the constraint exercised by it. Typically, the larger the forcing structure, the more extensive the behavioral constraint one would expect. The constraint exercised tends to be that of keeping more elements fixed in value, suggesting an explanation of why some nets have cycles in which many elements do not change state and are stable under perturbation.

The existence of large forcing structures would be promoted by higher densities of elements in the functional ensemble forcible on one or more inputs. As developed in Chapter 4, among the 2^{2^k} mappings possible for a given connectance, the fraction of mappings forcible on at least one one input declines quickly as k increases from 1, where all mappings are forcible. On this basis, one would expect the lower connectance nets (including the k = 1 case) to show larger forcing structures and the behaviors they promote. This simple inference is complicated by the fact that the existence of forcible inputs on an element does not ensure that it is a member of a forcing structure. Recall that forcing structures are chains of pairwise forcing elements. Pairwise forcing requires more than the simple existence of an element that *can be forced*. Also needed, upstream of the first element, and connected to it by a forcible input, is another forcible element, whose forced value in turn forces the first element. Therefore, the existence of large forcing structures must be argued separately, or the effect of increased forcibility assessed separately. Drawing on probabilistic graph theoretic results, Kauffman (1974, p. 83) argues that if the mean number of forcing inputs per element is two or more, typical nets have larger forcing structures. Further, if 60% or more elements are forcible on one or more inputs, the net has the behavioral characteristic of the k = 2 fully random net (Kauffman, 1974, p. 177). However, on this last point, it is not really clear to us where the 60% figure comes from. Nor is it clear that Kauffman intends it as numerically definitive. Since it is, or could be, a useful guide, it should be verified directly. Indeed, the relationship between forcibility and behavior deserves to be studied in more detail parametrically.

Kauffman's work to this point suggests the qualitative conclusion that nets rich in forcible connections, but deliberately constructed to avoid pairwise forcing connections, and hence forcing structures, *can* show short cycles (although perhaps not much cyclic stability). Hence large forcing structures are sufficient, but not necessary for short cycles. Hence it is possible that other small-scale properties may be involved in "desirable" large-scale behavior.

5.3.3 Kauffman Nets and Extended Threshold

As suggested above, other "small-scale" properties besides forcibility may be implicated in tractable net behavior. One might ask about alternative functional ensembles in anticipation of requirements imposed by modeling circumstances. For example, it might be that in some real-world situation, achieving a given level of forcing connections could be more difficult than achieving a given level of some other control variable. We now take up such an "other" control variable: extended threshold.

Extended threshold is defined in Chapter 4. To review briefly, a given mapping has an extended threshold of, say, 3, if a constant output is guaranteed provided that the mapping's input lines are carrying three or more 0's, or three or more 1's; constancy of output is not guaranteed otherwise.

Here we again come across the idea of mappings' output values being assured under certain circumstances. The guaranteeing circumstances now are not those of the structurally relevant notion of forcibility, but rather the more functionally based notion of a sufficiency of information or evidence. As developed in Sec. 4.6, there is a relationship between the two ideas. Here our main interest lies in how extended threshold itself affects large-scale behavior.

If a mapping has a low threshold, that term being understood to refer to *extended* threshold, its responsiveness to input changes is less. High threshold implies the possibility of greater responsiveness to changes in input conditions. From this, as well as from recalling the inverse relation between threshold and forcibility developed in Chapter 4, we might expect the behavior of low threshold mappings to resemble the fully random k = 2 net. These expectations in mind, let us consider an unpublished study in which we looked at behavior as a function of net size N, nominal connectance k, and maximum threshold L. Connectance values from 1 to 7 were used, for net sizes of 5, 10, 15, and 20. For N = 50, owing to the large aperture required for a sufficient likelihood of cycle disclosure, only k's of 1, 2, and 3 were run. In examining the effect of threshold we will consider only the N = 20 data, since that net size is the largest for which all k values were run.

We used *maximum* threshold because it is easier to construct flat maximum threshold ensembles, and because it also seems easier to conceive of real-world controls acting to constrain the *maximum* level of information required for consistent action than it is to believe that real world control would act to set *exact* levels of information--no more, no less--for consistent action.

The way in which mappings were chosen for a given value of L was as follows. Each of the mappings was taken in turn. With probability 0.5, the top or the bottom of the mapping was chosen. With probability 0.5, 0 or 1 was chosen. Then, from the top of the mapping down (or from the bottom up), 0's were filled in (or 1's filled in) for all mapping rows in which there were L or more 1's (or 0's). The remaining rows of the mapping were then filled in: 0's or 1's with probability 0.5. It can be seen that where L = k, one has a functional ensemble identical with that used in fully random Kauffman nets. (The data in Figs. 5.1 and 5.2 were taken from such points.)

Disclosure, Cycle, and Run-In Lengths. It is clear from Figs. 5.3 and 5.4 that maximum threshold controls typical cycle and run-in lengths. That is, the equiprobable choice of mappings from the set with a given

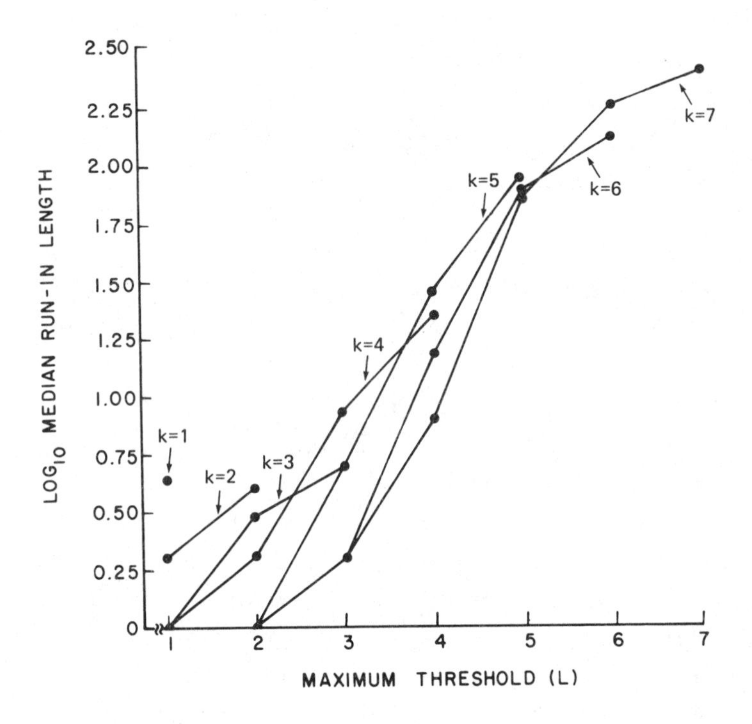

Fig. 5.3. Log median cycle length as a function of maximum extended threshold (L) and connectance (k) for N = 20 Kauffman nets (trivial nets excluded). One hundred trajectories were initiated for each point. One hundred disclosures were obtained for each, with the following exceptions: k = 5, L = 7: 99; k = 7, L = 4: 98; k = 7, L = 5: 99; k = 7, L = 6: 99. Original data.

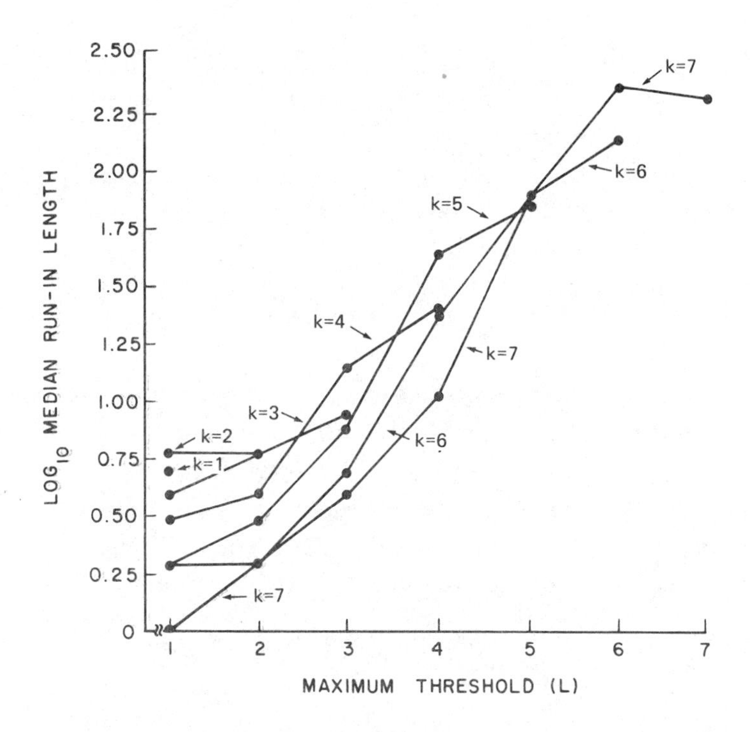

Fig. 5.4. Log median run-in length as a function of maximum extended threshold (L) and connectance (k) for N = 20 nets (trivial maps excluded). See legend of Fig. 5.3 for sample sizes. Original data.

maximum threshold sharply affects the size of cycles and transients obtained on random starts in the resulting nets' behavior spaces.

The relationships, for fixed connectance, between maximum threshold and both cycle and run-in length are quite similar. For the most part, as expected, increases in L increase both cycle and run-in length, on a rough ogive. The steepness of the ogive increases with the number of connections. Plotting disclosure length gives a display very much like that for cycle length, displaced upward.

Cycle Stability and Threshold. Figure 5.5 shows that, as expected, increasing the maximum threshold of a net largely decreases the stability of the net under unit perturbation. The observed exceptions are for extreme values of L in the case of k = 6 and 7. Remarks made regarding sampling error in Fig. 5.2 apply here as well.

5.3.4 Location and Distance Between States in Kauffman Nets

Locating net states by their density of 1's, and characterizing distance between net states by the density of element-state changes between them, have been discussed in Sec. 5.2.6. This section, which uses Gelfand and Walker (1982), considers the average location of and average distance between adjacent trajectory states in

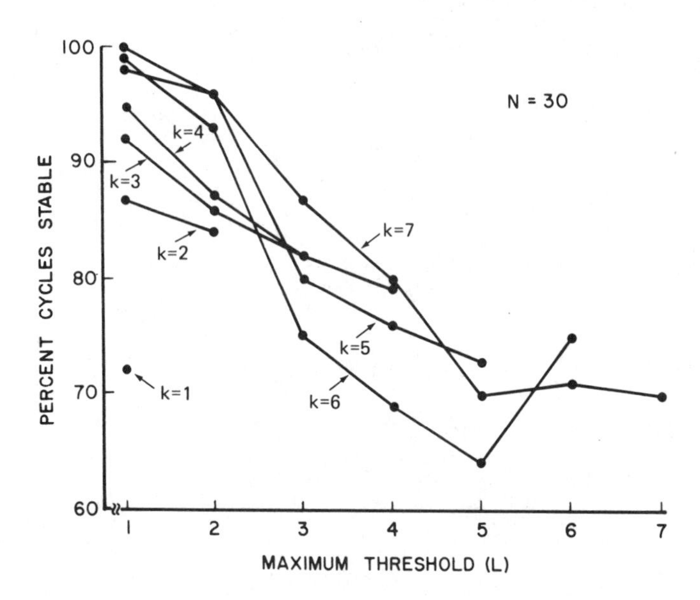

Fig. 5.5. Stability of cycles as a function of maximum threshold (L) under unit displacement. (Trivial maps excluded.) Legend of Fig. 5.2 applies as to sampling technique. For sample sizes, refer to legend of Fig. 5.3. Original data.

Kauffman nets employing four functional ensembles following theory described in Sec. 4.7. We therefore have nets in which sampling of inputs is done without replacement. The four functional ensembles are: fully random (case I); and asymmetric versions each of internal homogeneity (case II), forcibility (case III), and extended threshold (case IV).

Since the theory developed in Sec. 4.7 yields expressions difficult to evaluate where $k > 1$, simulation was used. For nets of size $N = 12, 36, 108$; for $k = 1, 2, 3, 5, 7$; for initial state densities of 1's of 0.25, 0.5, 0.75; and for each of the four cases, 100 nets were observed from $T = 1$ to 25 (occasionally to 10 where computing costs were high). The mean and standard deviation of the density of 1's in each net state and the density of changes between adjacent states of each T were calculated.

The "large net" theory of Sec. 4.7 gives simple expressions in all four cases where $k = 1$. We first take up the simulation results in these conditions to buttress the theory. Figure 5.6 illustrates a typical simulation in each of the four cases. The "theoretical" curves are denoted by "Th." The graphs demonstrate rather clearly that with increasing net size the simulated results--the real nets' behavior--tends to the theoretical prediction. The suggestion is strong that for very large nets the theory applies quite well. Recall that where $k = 1$, with-replacement and without-replacement sampling of inputs are equivalent sampling schemes, so that in particular, case I results correspond to Kauffman's in similar parametric conditions. In Fig. 5.6 and succeeding figures, nomenclature in the graphs is that used in Sec. 4.7. Figs. 5.7 through 5.10 illustrate typical simulations in each of case I to IV with $k > 1$.

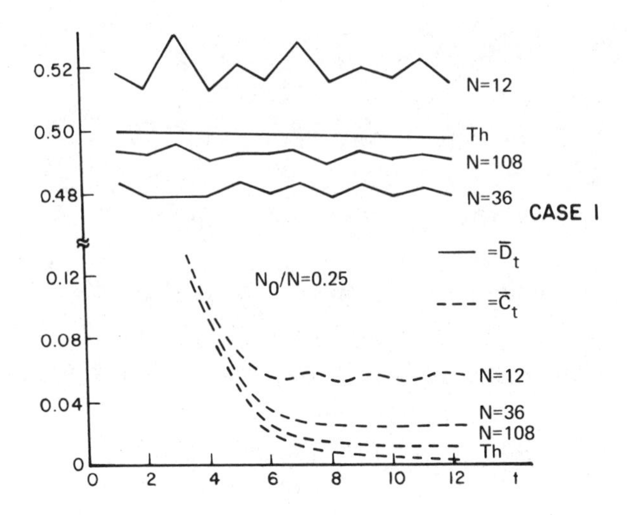

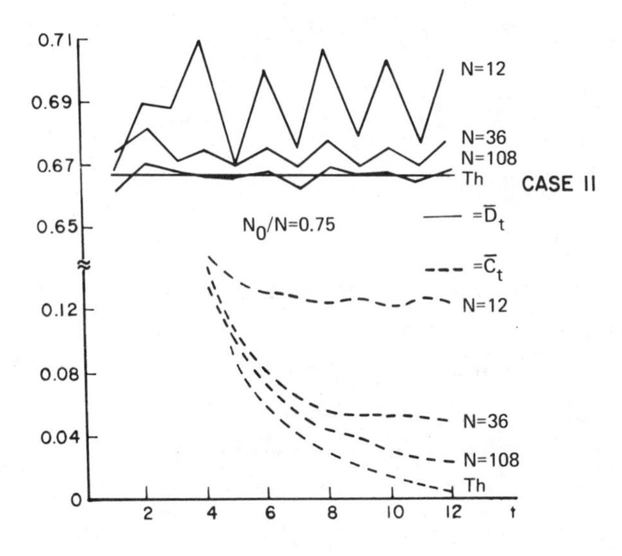

Fig. 5.6. Observed results and theoretical predictions of density of 1's and density of changes in k = 1 Kauffman nets over time. Each plotted empirical point is the mean of 100 observations at that time. Cases I, II, III, and IV. From Gelfand and Walker, 1982.

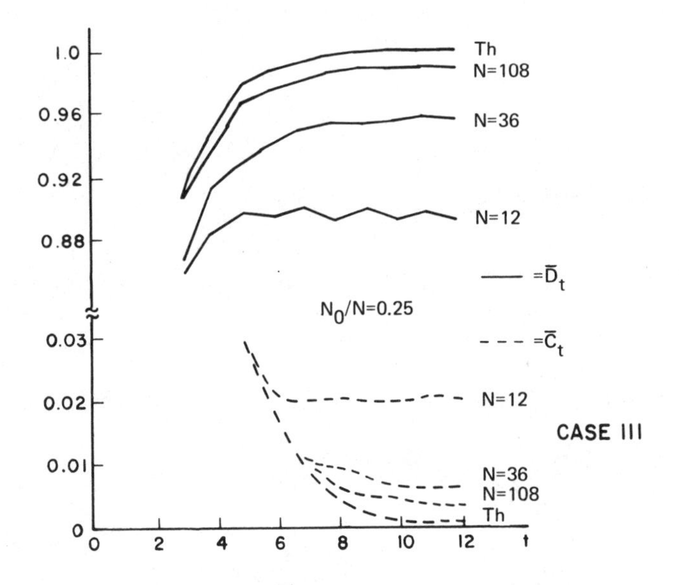

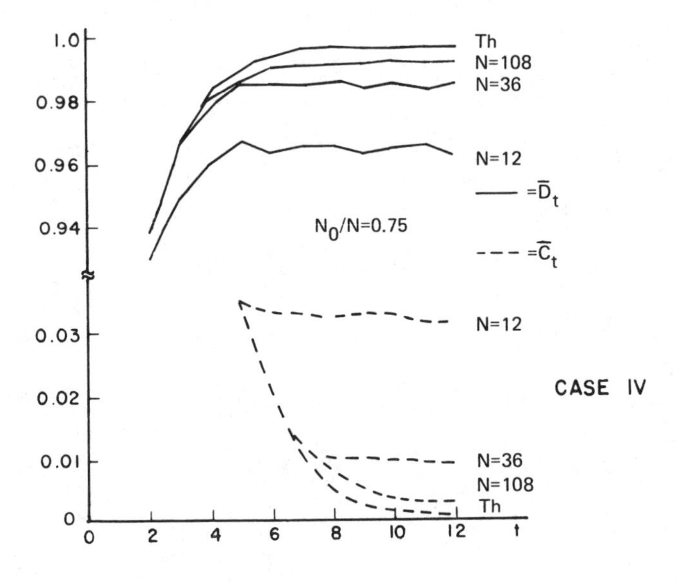

Fig. 5.6. Continued.

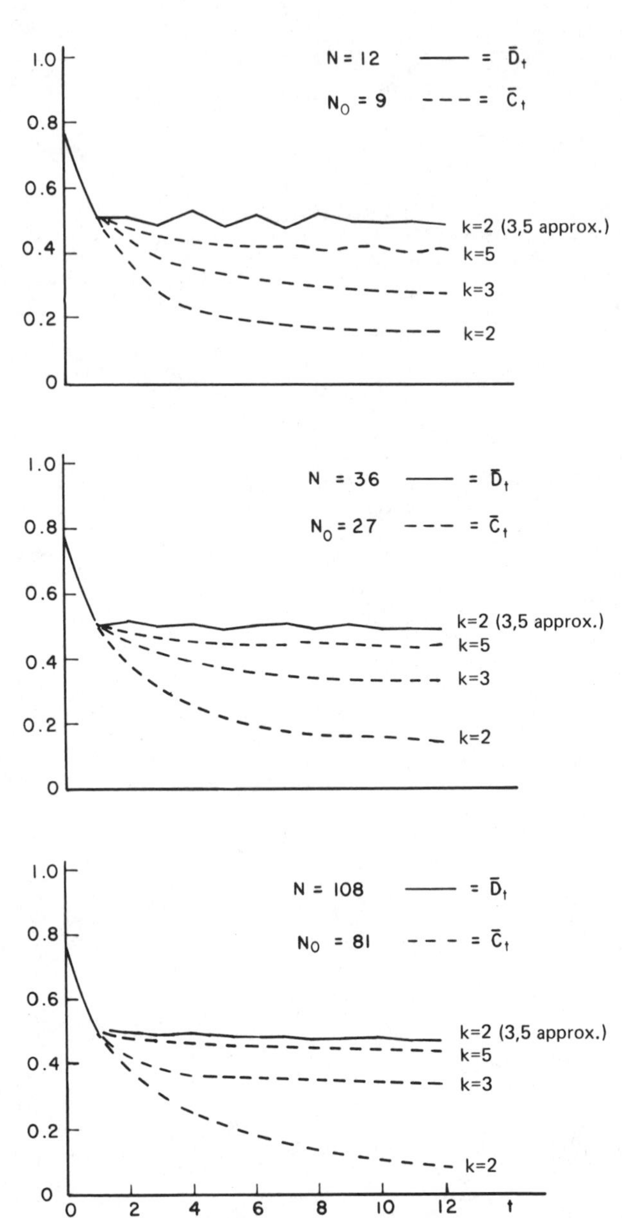

Fig. 5.7. Case I, fully random nets. Points plotted are means of 100 observations. From Gelfand and Walker, 1982.

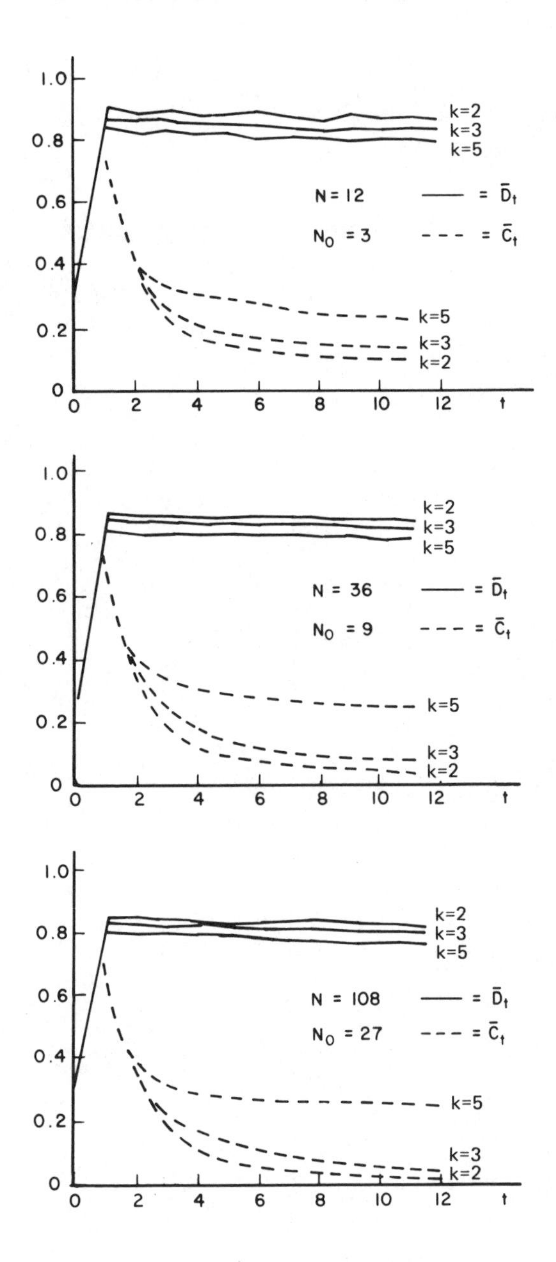

Fig. 5.8. Case II, asymmetric internal homogeneity. For level of control, see the text. Points plotted are means of 100 observations. From Gelfand and Walker, 1982.

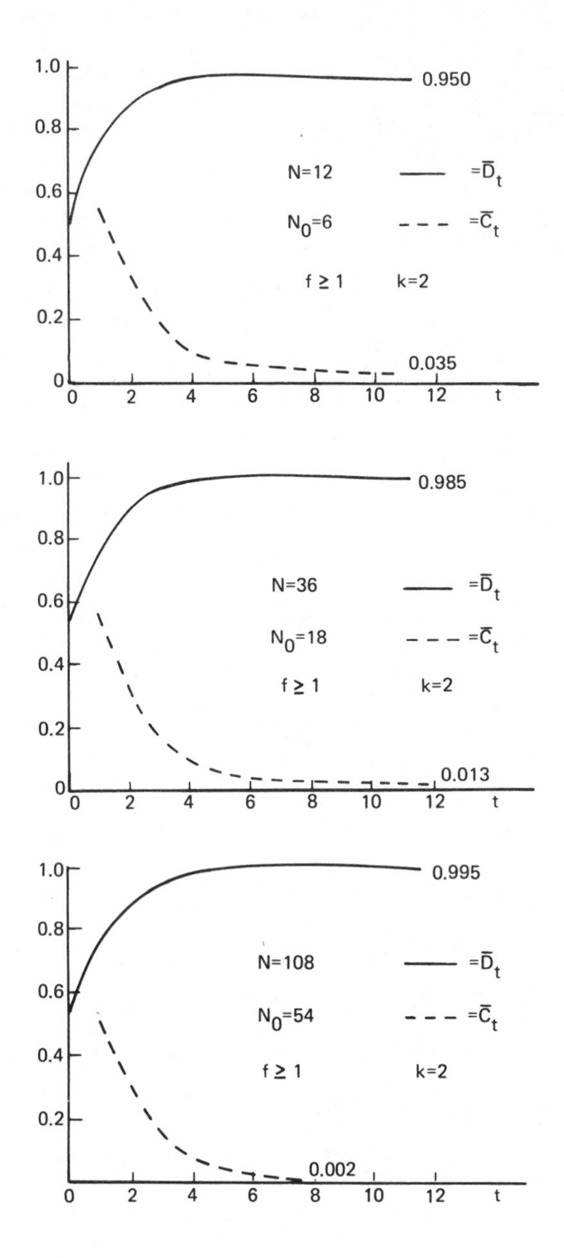

Fig. 5.9. Case III, asymmetric forcibility. For level of control, see the text. Points plotted are means of 100 observations. From Gelfand and Walker, 1982.

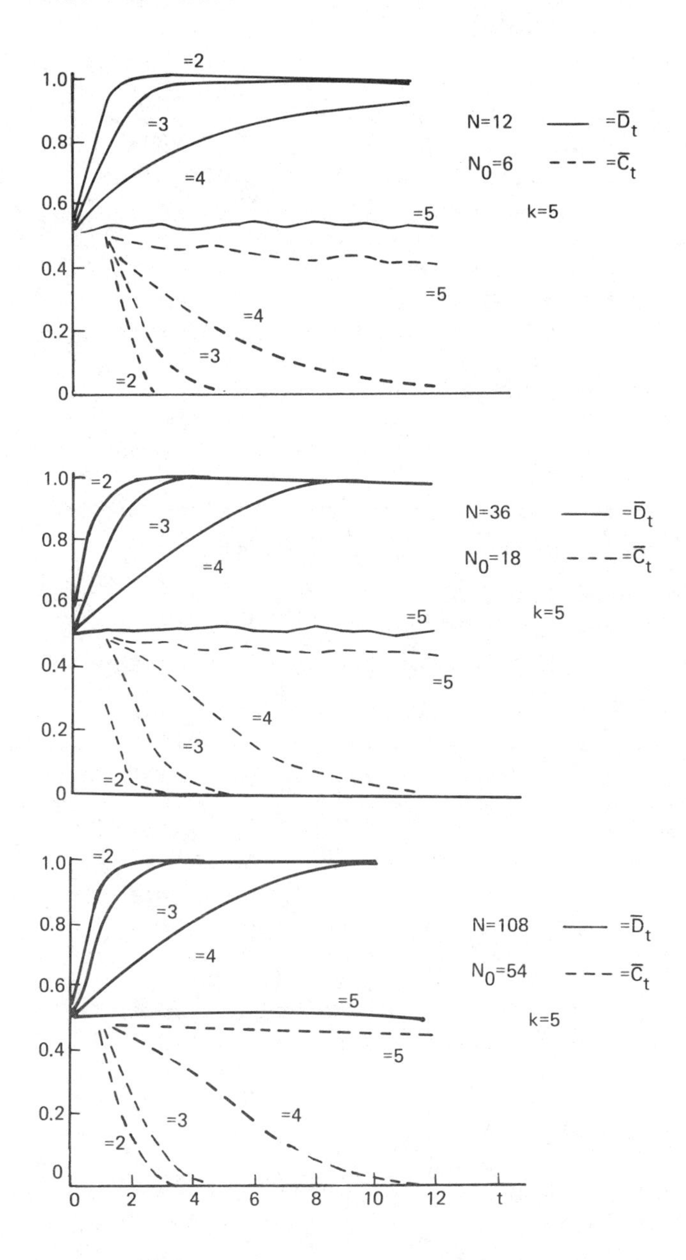

Fig. 5.10. Case IV, asymmetric extended threshold. Levels of control are indicated. Points plotted are means of 100 observations. From Gelfand and Walker, 1982.

Case I: Fully Random. Figure 5.7 applies. As expected, the density of 1's centers about 1/2 regardless of k and becomes more stable as net size increases. The distance between net states depends on connectance in a striking way. A dichotomy is suggested. When connectance is 2, the average distance between net states tends to the respective k = 1 case, while for connectance at least 5 (and to some extent for k = 3), average distance tends to the respective k = N (the fully connected case).

Case II: Asymmetric Internal Homogeneity. Figure 5.8 applies. The control used is asymmetric internal homogeneity, its level at least $(3/4)2^k$, using the truncated geometric model (successively higher levels of IH are one-half as likely to be obtained). As expected, the density of 1's at k = 2 is approximately 5/6, and with increasing connectance decreases toward its limit of 3/4. Increasing net size reduces the variability in density of 1's. For average distance, again a dichotomy is suggested. For k = 2 or less, average distance tends to the respective k = 1 case, while for k = 5 or more, average distance tends to the k = N (fully connected) case, where further with N large, the density of changes tends to the predicted 3/8. Note that in cases I and II, on average the net immediately reaches and operates at a terminal level of density of 1's. This is not so in cases III and IV.

Case III: Asymmetric Forcibility. Figure 5.9 applies. Since forcibility at least 1 is already a strong control, we do not consider higher levels. We set the probability of 1's in the mapping, given an achieved level of forcibility equal to the minimum, as 1/2. Since k does not enter into the theoretical prediction for density of 1's, and since the density of changes does not change much with k, we graph only the k = 2 results. As expected, the average density of 1's tends toward 1, and the density of changes tends toward zero as time increases, and even more so with increasing net size. For k sufficiently large with the level of control at 1, the density of changes would approach 3/8.

Case IV: Asymmetric Extended Threshold. Figure 5.10 applies, with connectance of 5 and the probability of a 1 being assigned in a mapping, given the maximum threshold value asserted by the control level, as 1/2. The graphs show that control levels less than 4 are quite strong, and are stronger as the control level decreases: the average density of 1's tends to 1, and the density of changes tends to zero very rapidly. At k = 7, a control level of 5 or less provides similar strong control for example. Generally, unless the control level is equal or perhaps nearly equal to k, this degree of control will occur.

References

Ashby, W. R. (1960). *Design for a Brain* (2nd ed.). New York: Wiley.

Fitzhugh, H. S. (1963). Some considerations of polystable systems. *IEEE Transactions on Military Electronics, Vol. Mil-7*, 1-8.

Gelfand, A. E. and Walker, C. C. (1977). The distribution of cycle lengths in a class of complex systems. *International Journal of General Systems. 4*, 30-45.

Gelfand, A. E. and Walker, C. C. (1982). On the character of and distance between states in a binary switching net. *Biological Cybernetics, 43*, 79-86.

Holland, J. H. (1960). Cycles in logical nets. *Journal of the Franklin Institute, 270*, 202-226.

Kauffman, S. A. (1969). Metabolic stability of epigenesis in randomly constructed genetic nets. *Journal of Theoretical Biology, 22,* 437-467.

Kauffman, S. A. (1970). The organization of cellular genetic control systems. *Mathematics in the Life Sciences, 3,* 63-116.

Kauffman, S. A. (1974). The large scale structure and dynamics of gene control circuits: An ensemble approach. *Journal of Theoretical Biology, 44,* 167-190.

Rubin, H. and Sitgreaves, R. (1954). Probability distributions related to random transformations on a finite set. Technical Report 19A, Applied Mathematics and Statistics Laboratory, Stanford University.

Sherlock, R. (1979). Analysis of the behavior of Kauffman binary networks--I. State space descriptions and the distribution of limit cycle lengths. *Bulletin of Mathematical Biology, 41,* 687-705.

Sloane, N. J. H. (1967). *Lengths of Cycle Time in Random Neural Networks.* Ithaca, N.Y.: Cornell University Press.

Walker, C. C. (1965). A study of a family of complex systems: An approach to the study of organisms' behavior. Technical Report 5, AF Grant 7-64, Electrical Engineering Research Laboratory, University of Illinois, Urbana.

Walker, C. C. (1971a). Behavior of a class of complex systems: The effect of system size on properties of terminal cycles. *Journal of Cybernetics, 1,* 55-67.

Walker, C. C. (1971b). Stability of the steady-state in a class of complex binary nets--a working paper. *Proceedings of the 4th Hawaii International Conference on Systems Science.* North Hollywood, Calif.: Western Periodicals Co., pp. 701-703.

Walker, C. C. (1973). Predictability of transient and steady-state behavior in a class of complex binary nets. *IEEE Transactions on Systems, Man and Cybernetics, SMC-3,* 433-436.

Walker, C. C. (1976). Cycle lengths of Ashby nets, empirical sampling for net sizes 4 through 17. Report 76-1, Complex Systems Research Facility, Department of Industrial Administration, University of Connecticut, Storrs.

Walker, C. C. (1978). Predicting the behavioral effects of system size in a class of complex binary nets. In R. Trappl, G. Klir, and L. Riccardi (Eds.), *Progress in Cybernetics and Systems Research*. Washington, D.C.: Hemisphere, 43-47.

Walker, C. C. and Aadryan, A. A. (1971). Amount of computation preceding externally detectable steady-state behavior in a class of complex systems. *International Journal of Biomedical Computing*, *2*, 86-94.

Walker, C. C. and Ashby, W. R. (1966). On temporal characteristics of behavior in certain complex systems. *Kybernetik*, *3*, 100-108.

6

Interpretation

6.1 Our General Modeling Domain

We return to the argument advanced in Chapter 1--that simple switching nets can be theoretically attractive examples of complicated systems. The present chapter amplifies this proposition using examples first described in Sec. 3.5, with additional material emphasizing the interpretation of ideas presented in Chapters 4 and 5.

As before, let us distinguish between static and dynamic net descriptions. By its static description we mean what is required to specify a net. By its dynamic description, we mean what the net does over time. To see our nets as models we will be required to furnish linkages between the nets and the real world in both descriptive areas.

It seems clear that the primary linkage motivating net based modeling lies in the area of static description. The real world appears to present a variety of situations in which numerous active points are linked together by transaction paths, and where the activity in the points is a response to incoming transactions. The face validity of network models, as *nets*, appears potentially high in many specific areas of organized complexity. It is in setting out the details of the nets where most of the questions of static resemblance arise. For example: Is the real world "truly" binary? Does it operate in clocked time? and so forth. Where possible in what follows we shall at least sketch arguments for detailed static correspondence between the net models offered and the real world being modeled. Where good evidence does not exist for all particulars of detail, we remind the reader that this fact does not, of itself, necessarily mean that the model is useless, particularly if significant behavioral fit exits.

Linkage between characteristic net behavior and real behavior can be much less obvious. For example, what can a net's cycle refer to in the real world? Stuart Kauffman's work has been important in making correspondence between net dynamics and real system activity interpretable. We shall take up these specific linkages as particular models are discussed below. We remind the reader that the ensembles we use are linked not to nets, but to the modeling context.

We recall that network models have been discussed in three specific substantive domains: biology, management, and advertising. We review these contexts and then take up the models and their implications.

6.2 A Biological Modeling Context

In modeling the genetic control system, Kauffman sets out his modeling context as follows:

> The continuing success of molecular biology in discovering the mechanisms controlling the expression of individual or small groups of genes is bringing to the foreground questions about the large scale organization and dynamic behavior of cellular control systems,

and how best to discover their nature. Current estimates of the number of genes, both control and structural, in a higher metazoan, range from 40,000 up to about 1,000,000. The uncertainty reflects in part the difficulty in assessing the role and extent of redundant DNA. Although it may turn out that these genes are organized into very simple control circuits, as suggested by Ohno (1971), that possibility is no certainty. Biologists should consider whether there may be some systematic way to use what is known to generate hypotheses about the likely organization of large systems of genes, how much of the cellular control system we can reasonably expect to know, and what might be explained with what we expect to know.

It is reasonable to expect continuing discovery of the major kinds of molecules playing control roles, their mechanisms of action, and of small scale, local properties of the organization of these molecular processes into control systems. In particular, it is proving possible to discover which molecular processes directly control a given process--the immediate control connections in a system; and how variations in the controlling variables affect the controlled process. Where these processes are grouped into reasonably simple intermediate sized control circuits, the properties of those circuits should be discoverable. This is already happening with bacteriophage lambda. However, it seems unreasonable to expect to discover virtually all the control relations among 40,000 or 1,000,000 genes. Therefore, we should consider ways to construct an adequate picture of the architecture of cell control systems whose full details may never be adequately known. In addition, incomplete knowledge of those control systems poses the critical problem that there are likely to be dynamic properties of central biological importance which depend in some way on large portions or on the whole organization of the control system, not on small isolatable fragments of it. A number of well known cellular dynamic properties probably reflect the overall organization of cellular control systems. Among these are: (1) the pattern of gene activities corresponding to any one cell type, in any organism, must be restricted to a relatively small number of combinations of gene activity through which the cell "modulates" its ongoing activities; (2) any organism possesses a particular number of stable distinct cell types; (3) during ontogeny, any cell type differentiates directly into rather few other cell types, although it may eventually differentiate into many by repeated branching differentiations. Large scale properties such as these presumably reflect and imply something about the overall architecture of cellular control systems whose design we wish to learn. If there are large scale dynamic properties of interest depending upon control systems whose full details are, and may remain, unknown, it is appropriate to assess how we can begin to link them to the kinds of small scale properties we expect to know.

One approach is to characterize any known small scale properties of the organization of cellular control systems, such as specifying the typical number of variables controlling any process and specifying the ways variations in the controlling processes affect the controlled processes. Specification of such small scale, local properties should be useful in two ways: (1) the local properties form the basis for hypotheses about the organization of larger control circuits; (2) the implications of the small scale properties for the large scale dynamic behavior of cellular control systems can be assessed. Systematic use of such local characteristics for both these purposes can be made by constructing a set of all the possible large control systems, each member of which is built using only those small scale properties. This set, or ensemble, represents the class of hypotheses about the total architecture of cellular control systems implied by known small scale properties of the organization. Examination of the typical, or average "wiring diagram" of ensemble members will allow hypotheses about the most probable kinds of intermediate and large control circuits which may be expected from small scale properties we already know. Examination of the typical large scale dynamic behaviors of the ensemble members will allow us to assess the most probable large scale behaviors of cellular control systems having the known small scale properties. The primary purpose in characterizing small scale properties of cell control systems and constructing

> an ensemble of possible control systems is to examine the implications of known small scale features for probable large scale properties, rather than directly to help learn about molecular mechanisms or the small scale properties themselves. [Quoted with permission of the author and publisher from Kauffman, 1974, p. 174ff. Copyright: Academic Press, Inc. (London) Ltd.]

Kauffman's purpose is explanation of gene-produced dynamics in a domain where certain small-scale properties of gene systems appear to apply generally but where large-scale static properties are unknown. His object is to specify how variations in the small-scale properties affects large-scale behavior. If certain small-scale properties typically produce systems showing good large-scale behavioral (or structural) real-world fit, and other small-scale properties do not, given the assumed uncertainty regarding large-scale specifics, the former properties provide a better explanation of system behavior than do the latter.

In a biological area an evolutionary argument is also appropriate. If certain small-scale properties on their own commonly give rise to large-scale dynamics that promote biological survival, aggregates that make use of such local properties might well enjoy a survival advantage over aggregates whose survival is more contingent on large-scale properties that are specifically detailed. If such useful small-scale properties exist, we would expect them to be evident in at least some organisms, and to be of particular pertinence in those large-scale properties that occur in *different* organisms.

6.3 A Management Modeling Context

Kauffman does not go so far as to claim that it applies in his context, but it is conceivable that even if all the control relationships existing among real genes were known, the logical structure of the resulting model could be so overwhelming as to provide little understanding of behavioral dynamics, and hence little theoretical gain. Kauffman's position is more conservative: We may never be able to know all genetic details. In our managerial model we are still more pragmatic: Even if a manager could come to understand, in detail, the organization being managed, surely it can happen in practice that constraints on time and resources may require the manager to exercise control under conditions in which use of that detailed organizational knowledge is not possible. Our context sees the manager as able to specify certain local features of organizational control systems, but unable to be certain of, or perhaps, to use knowledge of, the whole fabric of their action or interconnection. The "typical" manager's practically limited rationality motivates our use of ensembles of structures and functions.

The main features of our managerial model are doubtless predictable by this point. Focusing on the problems of controlling repetitive organizational behavior, we consider an organization's control system as a collection of nodes in which information is processed, connected by paths through which information is passed between nodes. We approximate these aggregates by switching nets. Our goals are more metatheoretical than Kauffman's. Although some of our interpretation is intended as serious substantive comment, more importantly we intend to use the model to illustrate how theory building might plausibly proceed given our modeling context.

To be consistent with our modeling context, the control available to the manager should not presume controlling action to involve specific, detailed interventions. In our scheme the manager is allowed only broad control of certain small-scale properties, in particular (1) the number of inputs to nodes in the net, and (2) the *classes* of functions from which the net elements' mappings are selected. We later argue that these (equivalence) classes, the latter defined over net mappings, can be seen as managerial strategies or managerial styles, and that suggestive parallels can be drawn between them and some recognized managerial techniques.

6.4 A Market Modeling Context

As a third example, we examine an activity, advertising, in a context which resembles that of the managerial model described above. Our purpose in this modeling is largely metatheoretical. Hence a fairly strong flavor of "stone soup" theorizing is to be expected. We intend to exhibit some of the possibilities inherent in a net-based ensemble framework, interpreted in advertising terms.

To focus the modeling, we will consider advertising for a very particular purpose: to influence people in their choice of a real estate broker. Why this narrow domain? It appears to fit our modeling technique quite well. Within the last decade it has become clear that the home-buying process cannot be well described by conventional models (Hempel, 1970). Home buying appears, rather, to be an exceedingly complex many-variable, time-dependent process. It is not beyond hope, however, that at least some facets of the process can be usefully examined separately. One such facet may be the word-of-mouth dynamics that influence prospective home buyers in their choice of a broker.

Word-of-mouth is an important variable in home buying. Word-of-mouth has been cited (Hempel, 1969, p. 47) as a factor second only to specific property-oriented newspaper advertisements in influencing a buyer's choice of a particular brokerage firm. Moreover (Hempel, 1969, p. 73),

> The frequent use and apparent importance of word-of-mouth in the home buying process may be attributable to the convenience and subtle influence associated with advice obtained during the ordinary routine of social interaction. Whatever the explanation for this pattern of exposure and influence, the broker must clearly be concerned with the informal communication channels that complement and sometimes dominate the commercial channels with which he is likely to be more familiar.

Since word-of-mouth interactions among people over time is a complicated phenomenon involving individuals whose intercommunication patterns are intricate and generally unknown, and whose responses to communications differ and are not known, it is an area where ensemble modeling may be useful. We therefore propose to use such an approach in modeling the time course of word-of-mouth opinions toward specific brokers. We use the data discussed in Chapter 5 to suggest how the commercial channels with which the broker is more familiar might be expected to affect the word-of-mouth-produced dispositions the broker might otherwise find relatively difficult to influence.

As mentioned already, we assume that in any influence the broker exerts, detailed restructuring of the influence group interaction pattern and person-by-person changes in information usage are impossible. As in the managerial model, inability to affect relevant details of the situation directly justify the ensemble approach. The influence the broker is assumed to exert, through advertising, is interpreted as consisting of restrictions on the populations of mappings from which net elements are assigned. We suggest that some of the formal controls discussed in Chapter 4 can be interpreted as types of advertising message control.

6.5 Kauffman's Genetic Model

Starting with this section, the discussion will deal more completely with the models of interest. The first considered is Kauffman's model of the genetic system, proposed in 1969 and elaborated on in a series of papers (Kauffman 1969, 1970, 1974, for example). These original writings should be consulted for biological details. Our interest is in explicating the model. We shall describe (from a layman's perspective) the links between the model and the "real" world, attempting to specify what purpose the specific links serve.

Static Linkages

Biological Observation	Consequences for a Model
A living thing is a complex web of interactions between thousands or millions of chemical species.	A network is an appropriate generic model for systems which set up or control biological systems.
A gene specifies a protein, and that protein can control the rate of output of other genes. Gene products "derepress" and "repress" other genes' products.	Genes' effects are determinant and can be chained together; an on-off Boolean model gene is not unrealistic or continuous gene action.
Cross-reactions between real genes do not appear to be constrained in any simple way.	The model gene net can have a complicated feedback structure.

Biological Observation	Consequences for a Model
A given real gene's products affect specific other genes' actions.	The net structure, once given, should remain fixed.
Genetic systems appear to operate without necessarily requiring externally-given information.	The net can be autonomous.
Living genetic reaction nets are unlikely to be composed of genes all of which react the same way to their controlling genes.	The nets should be functionally heterogeneous.
Remembering that real gene systems are separately and individually located in cells, a description of the cell's biochemical status at any given time is provided by the status of its collection of genes as regards their individual products at that time.	The net state at a given time signifies the gene systems', and hence the cell's, biochemical status at that time.
The number of genes may well be an important biological variable.	If N is the number of model genes in a model gene system, let N be a primary modeling variable.

The following links appear to be motivated both by considerations of convenience in the model and by a lack of strong disqualifying biological evidence.

At least some genetic actions can be very roughly simultaneous (e.g., in cell division). Appeal might be made to similar reaction latencies for similar biochemical reactions.	The net operates in discrete time, and the output of a model gene at time t is felt by all controlled genes at time t + 1.

Ensemble Linkages

Context	Model Ensemble
The specific large-scale structures in genetic systems are unknown.	Inputs are joined to outputting elements at random with replacement. (Benchmark ensemble.)
The character of gene actions in place in genetic systems is not known.	Mappings are assigned at random with replacement to net elements. (Benchmark ensemble.)
The circumstances in which real genetic systems "start" are not known. At least some aspects of the behavior of the model system may be discovered by starting it quite arbitrarily in its behavior space.	Initial states for net trajectories are selected at random with replacement from the set of net states. This is true except where starting conditions do appear to be defined (e.g., in testing the stability of the net under perturbation). This ensemble appears to be determined more by the technical necessity for a starting state than by clear considerations of linkage.

Behavioral or Dynamic Linkages

Biological Observation	Model Interpretation
Cell types are distinctive, stable productions of genetic systems.	Cycles correspond to cell types.

Kauffman's key behavioral interpretation is the foregoing cycle-cell type link; his other behavioral interpretations are derivative. Since the cycle-cell type link is so important, let us expand its interpretation, drawing on Kauffman (1969, 1974). First, a cell type appears sensibly conceived of as one of a real gene system's several distinct steady states, perhaps one that shows oscillations in some chemical species. Second, large numbers of cell types show cyclic behavior, marked clearly by cell division. Third, in cells that do not divide, some sort of continuous repetitive process appears implicated to replenish chemical products consumed in ongoing cell activities.

To summarize the above, each cell type is a relatively permanent status, produced by genetic activity, which must be a sharp restriction of biochemical activity to a very constrained subset of all conceivable productions of which the same gene system is capable. Thus, whatever the case may turn out to be for a strict cycle-cell identity, the behavioral restrictions which cell types embody, their relative permanence, and their distinctiveness one from another are features so clearly shown by net state cycles that *not* to pursue the possibilities of the correspondence would appear to be the more questionable alternative.

Behavioral Fit. Having specified in the form of a model those properties of the genetic system which largely are accepted, we can now consider those genetic properties that are yet to be explained. That is, what important biological properties might be related to the "knowns" of the model gene system? In the present case what seems least well understood, according to Kauffman's context, is primarily the developmental history of the gene system. Accordingly, Kauffman provided linkages between the model gene system and the biological dynamics of interest. He could then ask how well the model dynamics fit what is known about the behavior of the real organism. In summary, his findings are as follows. By constrained nets we mean low ($k < 4$) connectance, or nets with at least 60% of their elements having a forcible input.

Model	Biological Datum
In constrained nets of biological size, cycles are typically short. In unconstrained nets otherwise similar, cycles are on average much too long to be biologically meaningful.	In comparison with their potential behaviors, cell types must represent enormously constrained patterns of gene activity.
In constrained nets, the number of distinct cycles in the typical behavior spece increases approximately as the square root of N. In unconstrained nets the number increases with N.	The number of cell types in an organism increases as a fractional power of the number of genes.
In constrained nets, cycles are very stable under perturbation; not so for unconstrained nets.	Cell types show homeostasis.
In constrained nets, a net can be induced to shift under small perturbations to only a few neighboring cycles. This is not so for unconstrained nets.	In virtually all developing organisms, no cell type differentiates directly into more than five or six other cell types.

6.6 Critique of the Kauffman Genetic Model

Even though their individual fit may be only approximate, the simultaneous fit of several logically independent factors is strong evidence that the model should be taken seriously. As Kauffman (1974) points out, this is not a claim that cell control systems are random. What has been exhibited is a remarkable similarity between behaviors that certain small-scale properties typically provide, and behaviors that cell control systems actually show. It is conceivable, of course, that organisms have control systems which are precisely determined in the large. No amount of similarity between the real thing and what small-scale properties make possible proves that organisms *are* "small-scale" mechanisms. But the correspondence is strong enough to make it appropriate to ask not only how the working out of the small-scale properties in the large can be understood, but also how in the course of evolution the exploitation of small-scale properties' manifestly useful dynamics could have been avoided.

In explaining how small-scale properties work out in the large, it will be necessary to seek net-based mechanisms which have a high likelihood of producing the more specific cell type transitions seen in developing organisms. A better appreciation of details of typical behavior spaces (e.g., the spacing of cycles, stability contours, and what sorts of factors influence them) would be useful in that connection.

As to weak points in its static description: (1) Perhaps the most obvious weak point is the binary nature of the model. However, Kauffman (1974, p. 179) counters this objection, and even as admitted approximations, binary-state genes appear close enough to reality to be informative. (2) A net in which all elements have exactly k inputs is almost surely incorrect as a representative model of the gene control system. However, as related to theory the data derived from such nets show the importance of this variable and strongly suggest that gene, and perhaps other biological control systems, are biased toward small average connectances or sharp functional control, or both. It would be useful, of course, to verify the effect on behavior of connectance when it is interpreted in more biologically relevant ways (e.g., as the average number of inputs per element). (3) To us, as nonbiologists, the clocked-time assumption seems the least satisfactory static model descriptor. Further examination of both the biological fit of discrete time in the static model and the influence of this property on the behavior of nets appears important.

As to potential weaknesses in the ensemble linkages described above, one is particularly evident: the complete independence between the assignment of mappings to net elements and net structure. It seems clear that if biological evidence is available, or can be generated, regarding small-scale dependencies between function and structure, the behavioral influence of these additional properties should be examined closely. Such additional details make the model more complex, but may provide a means of "explaining" the more specific behaviors of these more realistic nets. It should be noted that Kauffman's (1974) discussion of forcing structures is aimed at just this sort of explanation. Another objection has to do with the use of bench-mark ensembles. Naturally, bench-mark ensembles can be objected to, in part or in whole, as far as their representative status is concerned. In the development of theory, they are useful as easily comprehended reference points, hence the name. In their given modeling contexts they also serve as theoretical statements until other considerations require their modification. We are not in a position to comment definitively on the validity of the ensembles. However, it does seem that from a representative point of view, one might question in particular the "with replacement" sampling of inputs. Such a sampling scheme says that one is taking it to be the case that the duplication of inputs is not less likely than the appearance of any other pairing of inputs.

Let us sum up. By small-scale mechanisms we mean devices which are constructed, were designed, or can be understood by the use only of small-scale properties. Kauffman's thesis is: (1) If real gene control systems are small-scale network mechanisms, they must be constrained systems. He gives two examples of controls powerful enough to do the constraining. Moreover, and more surprising, (2) real gene control systems may indeed be small-scale network mechanisms--organisms show a number of biologically significant large-scale behaviors that are produced by small-scale networks, and the constraints suggested by the model can be seen, to some degree in any case, in real gene control systems.

That gene control systems may be small-scale mechanisms is an important biological statement well worth pondering. To agree with it is to believe that one may *not* need to expect a basic orderliness to underlie all aspects of the intricate, recurring sequences and forms of biology. That the contrary expectation is part of the current biological paradigm may explain why Kauffman's model has been slow to affect the field, despite its elegance and explanatory scope. Recent developments in other areas of biology suggest that this predisposition may now be changing.

6.7 A Management Model

Our primary aim in this section is to illustrate how small-scale, network-based modeling methods might be used in the theory of management, especially in the management of complex organizations. Our focus is not so much on the model and its linkages, but more on the way the modeling approach might be used to formalize and extend some aspects of management theory. Except for the more speculative interpretations later in the section, the discussion is taken from Walker and Gelfand (1979) and Gelfand and Walker (1980), with some changes in interpretive emphasis.

In particular, there are two main points made. (1) Ensembles can be used to represent the perspective of a manager dealing with a complex organization. (2) It may be possible to formalize some management strategies or management styles rather directly, for example, as equivalence relations over control system unit functions.

These two points are developed in the modeling context discussed above using networks as the generic model type to approximate organizations whose control system structure and unit functions are fixed but unknown in their details. We first lay out the network analogy more exactly. Then we reintroduce four formal control variables discussed in Chapter 4. Knowing how these variables are related, from the combinatorial results developed in Chapter 4, allows us to suggest, from Chapter 5 data, how they will affect network behavior. Finally, we draw parallels between the formal controls and existing or potential real-world control strategies. The managerially relevant ideas we link to our formal controls are (1) an input side analog of the notion of span of control, (2) the classical exception principle, (3) the scalar principle of management, and (4) consensus management, or, more exactly, consensus-level management.

The Network Analogy. The network analogy appropriate here sees the switching net as a control system embedded in an organization, or only some part of it; there may be several control systems in one organization. The control system is not restricted to single-level activity but may include hierarchical aspects as well. The net elements correspond to points in the organization where control information is used. The use could be executive, evaluative, productive, and so on, and would depend on the particular activity being modeled. For example, the net elements could collectively represent a group of middle managers carrying out organizational routine, staff personnel reaching a decision, or foremen and their crews producing goods. The network structure represents the control relationships which exist among the elements, that is, the channels through which actual control information is passed in the system. The responses of a control point to the possible contingencies presented to it by its sources of information collectively are represented by the element mappings. If the organization to be modeled is a production process, net elements might represent machines or assembly points, each of which is capable of manipulating at most two distinct types of items or materials. In this illustration, a behavioral cycle models a product, the states preceding the cycle model the startup period, and the entire set of cycles gives the inventory of products the particular control structure and set of constituents is capable of producing.

Note that in these nets all elements' inputs are connected to element outputs. Net operation is not affected by external information. Therefore, if the model's static fit is at issue, the organizational systems modeled must be (1) effectively insulated from the environment, as perhaps by inventory policy in the case of a production system, or by high echelon planning, as in the case of a middle level management system; (2) tied to the environment by a relatively slow or infrequently acting control loop, as might be the case for stockholder influence in a corporate model; or (3) such that the environment itself is so intimately linked to the organization that it can be appropriately included in the net.

In our modeling context the network structure is taken to be fixed but unknown in detail. This presumption is surely descriptive of at least some aspects of management, particularly in complex organizations. As mentioned in the discussion of the context above, even though an organization's formal and informal organizational control structure might be fixed, and in principle, specifiable, the time required to know those details and to make a prediction of the effects of some proposed change might well exceed the time available. Therefore, both a complete theory of management, and the operating manager as well, would appear to need some way of predicting what behavior can be expected under conditions of structural and functional uncertainty. As we have argued, one method of accommodating that uncertainty is to study appropriate ensembles of structures and mappings. This method shifts analytic interest to the behavioral properties of the set of control structures generated. The data we cite and the applications we suggest refer to flat ensembles of structures and specified subensembles of functions. Of course, the generation of behavioral information on ensembles tailored appropriately to specific managerial circumstances would be required to bring this model closer to practical utility.

It may be that control systems in larger social and organizational domains can be usefully examined through network models. However, our interpretation will focus on what can be done within single organizations by varying strategies, or by maintaining different intensities of use of a strategy.

The Formal Controls. To review the formal concepts used in this section, first recall that the number of inputs which all elements in a given net have is one such control. The forcibility of a mapping is a count of the number of inputs which are forcible, that is, which can bring the mapping to a given value regardless of the other inputs. A forcible mapping is one that has at least one forcible input. The internal homogeneity of a mapping is the number of appearances of the more prevalent entry in the mapping table. The (extended) threshold of a mapping is the least number of inputs whose values need be specified to ensure the that mapping takes on a specified value.

Effects of the Formal Controls on Repetitive Network Behavior. From Chapter 4 combinatorics it is known that decreasing the number of inputs increases the density of logic functions with that number of inputs which are forcible. Increasing internal homogeneity increases the density of functions with that internal homogeneity which are forcible; and decreasing threshold increases the density of mappings with that threshold which are forcible.

From Chapter 5 data we know broadly that for small-scale Kauffman net modeling of repetitive system behavior, constraints on the nets are necessary to make average cycle length empirically reasonable if the nets are anything beyond small in size. More specifically, holding the number of inputs down, ensuring sufficiently high levels of forcibility, or reducing the maximum threshold allowed in nets all serve to reduce average cycle lengths to "empirical" values. Taking these "main effects" to apply generally--an assumption that certainly deserves direct verification--we can use the Chapter 4 combinatoric relationships to predict further conditions in which "reasonable" cycle lengths should be predominant. For example, increasing a net's average internal homogeneity should reduce expected cycle lengths owing to the positive relationship between internal homogeneity and forcibility. Similarly, if one holds the net's average threshold constant while changing average internal homogeneity, the average cycle length would be predicted to go up or down with the change in internal homogeneity levels.

The Formal Controls as Management Strategies. Can any sense be made of these formal concepts? We suggest that it is possible to do so. Our object in what follows is to build as convincing a case as we can for the usefulness of attempts to interpret these formal concepts. We think the specifics of the interpretations are less important at this time than the evidence, which we shall try to provide, that *some* interpretation, perhaps different from what we give below, will be useful in the theory and practice of management. In this interpretive effort, it is important to keep the modeling context in mind. Details of large-scale organizational structure and function are unknown. Given this, we ask: What managerially relevant concepts can be linked to the formal controls? And so interpreted, how might they be useful to the organizational theorist or to the operating manager?

Number of Inputs. Our interpretation of the number of inputs is straightforward. The manager manipulating the number of inputs in his or her control nets is managing "input span." This concept is clearly homologous, on the input side, to the idea of span of control or span of management (see, e.g., Koontz and O'Donnell, 1968, p. 241).

Threshold. In dealing with extended threshold, we observe that consensus, when there can be more or less of it, implies the existence of a metric on which the amount of evidence of disposition "for" or "against" something can be scaled. An extensive literature exists regarding consensus, especially as applicable in political and economic contexts. Unfortunately, aside from a common interest in the concept, this literature and our present interest differ, both in their basic models and in the behavior explained. One particular difference deserves emphasis. Here attention is focused not on exercising control by increasing agreement or disposition toward something, but by changing the levels *at which agreement is deemed to exist.* This is a very different orientation.

To return to the problem of scaling disposition "for" or "against," it can be seen that neither internal homogeneity nor forcibility provide a measure relating the amount of incoming evidence to action. Furthermore, consensus invokes the idea of a a level at or about which some course of action is taken or some condition is maintained; that is, consensus is interpretable as embodying a sufficiency condition for the undertaking of an action (or the setting of a condition).

These two ideas, scaling inputs in some way as to how much there is of some quantity in them, and the sufficiency condition, are inherently modeled by threshold functions. Conventional binary threshold functions (Dertouzos, 1965), however, typically interpret "0" as "off" and "1" as "on," and provide both sufficient and necessary conditions for setting output values. In addition, conventional threshold functions typically set the "on" condition for exceeding of the threshold level. In a managerial environment it might be useful to take no action given sufficient evidence, or to take an action given a substantial lack of something in the input. The notion of threshold formalized in Chapter 4 allows these possibilities. Thus, the manager seeking to control an organization by manipulating threshold levels will be said to be attempting to manage by consensus level (MBCL).

Note that in this interpretation, the total amount of disposition that affects any given control unit at any given time is simply the number of inputs showing that disposition at that time: for any given control unit, all its information sources are considered equal with respect to these dispositional "votes." Note also in this interpretation that 0's are considered equal to 1's in managerial significance. That is, we have an interpretation which is symmetric in input (and output) values.

Forcibility. To develop an interpretation of forcibility, consider a manager or consultant attempting to increase forcibility in an organization's control system. If for the sake of illustration the control net units are assumed to be occupied by people, the manager might address them as follows: "I want you to look carefully at your mission and potential sources of control information. Arrange your activities so as to increase the number of information sources whose orders individually warrant action, or inaction." So stated, this control style is perhaps harder to characterize than the previous two, but setting priorities on information sources seems intuitively to be an important part of its content. For that reason we call this managing style "management by (input) priority" (MBP).

This control style recalls the "scalar principle" of management (see Koontz and O'Donnell, 1968). The latter requires that for effective management each individual has a (single) boss. One difference in the two concepts is that our model views the organizational control system as allowing "horizontal" arrangements, that is, where information sources may be on the same level as the controlled unit. This follows from the fact that loops may exist in the control structure. The scalar principle, on the other hand, implies a consistent vertical boss-subordinate relation between connected controlled units. As to whether the interpretation offered here is therefore not representative, we remark that it may be questionable how often real control systems can be expected to have control structure which show mathematically pure hierarchies.

This is not to suggest that the scalar principle is necessarily one-dimensional in interpretation. The discussion above has emphasized the authority relation implicit in forcibility. It should also be clear that the scalar principle's requirement of a *single* boss can be understood as a requirement that connectance be set to 1. As we saw in Chapter 5, this connectance value is a strong behavioral constraint.

Internal Homogeneity. Now consider the same manager attempting to increase internal homogeneity: "Look carefully at all the control orders (the various input states) you can be confronted with. We know that no matter what the order, you can do only one of two things. I want you to try to arrange your responses to these orders so that one of them occurs much less than the other. Taking your mission into account, make up your own mind as to which that should be."

So stated, the manager's tactic appears to embody much of the familiar "exception principle" of management (Taylor, 1947). Taylor apparently did not define the concept rigorously. Instead, he gave examples of how it might be put into use: for example, by the hierarchical resolution of disputes, by reporting conventions, or by the delegation of duties (Taylor, 1947, pp. 109, 126). Nevertheless, Taylor apparently felt that management by exception was indeed a management *principle*. That is, he seemed to think that there existed a conceptual core, the principle itself, which could be dealt with appropriately by an operating manager, even if only intuitively.

The elucidation of that core he was apparently willing to leave to others--perhaps he thought its content was obvious. In our present context, seeking to draw into the model nothing more than what it has now, we therefore say that the manager seeking to vary internal homogeneity is attempting to modify his or her use of management by exception (MBE). There seems little alternative: the parallel is intuitively compelling in the sense that it appears contained in Taylor's ideas.

Given our reduced conceptual circumstances, arbitrarily structured nets, symmetric interpretation of 0's and 1's and so forth, it may be asked why we are even trying to interpret an idea as rich as MBE. With no explicit hierarchical control structure in place, how can the model accommodate MBE as conventionally understood? What *is* an "exception" in this model? Our point is that if it is possible to interpret MBE in our model, it should be done. An adequate theory of management should be able to handle simple situations as well as those that are complicated. Getting at core meanings is the point of theory building in the first place. It is not an important problem that MBE may appear simplistic here.

A more pertinent objection is that the idea of relative infrequency invoked here is implicitly calculated on the assumption that all control orders are equally likely. Taylor perhaps would have preferred an effort to represent more realistically the likelihood of control orders. The point to be made here is that under this alternative interpretation, MBE would be represented by something like internal homogeneity.

In a similar way, if better model fit were obtained with a nonsymmetric interpretation of 0's and 1's, the informal statement of MBE might appropriately change: "Look carefully at all your control orders. I want you to try to arrange your responses to these orders so as to take action as seldom as possible." If anything, this sounds *more* like what Taylor had in mind, and clearly, the manager is not attempting to manipulate internal homogeneity, strictly construed. On the other hand, he is attempting a manipulation of a variable which closely resembles internal homogeneity, given a decisive disinterest in "inaction" as an outcome. That is, internal homogeneity as defined is a measure of salience or predominance of entries in a table--take your pick as to which dominates. A simple count of the number of 1's in a table is also a measure of the dominance, but one that is concerned exclusively with one type of table entry.

At issue is whether or not we are claiming that internal homogeneity *is* MBE. We are not. We are claiming that *in some models* internal homogeneity is MBE, and we are suggesting that in other models, MBE may be represented by equivalence relations similar to internal homogeneity.

Some of our concern with internal homogeneity results from the simple fact that more is known at the present time about how internal homogeneity and other 0-1 symmetric measures of mappings affect network behavior. In Sec. 6.8 we consider some nonsymmetric cases. Immediately below we begin the task of explaining how the linkages suggested for the present model and the data described might speak to questions of organizational theory and managerial utility.

Some Implications for Organizational Theory. The suggested interpretations and the evidence available concerning the behavior of large nets allow us to address the question of the general nature of management by exception, and to understand in particular why MBE might be expected to be quite broadly useful. MBE, at appropriate levels of intensity, provides relatively high internal homogeneity, and therefore tends to maintain suitably high densities of forcible mappings in the organization's control nets. This, in turn, promotes usable, stable behaviors in a wide variety of structural circumstances.

Further, it can be seen that management by exception, management by priority, and management by consensus level are distinct managerial strategies, but in specified settings they can have a specifiable relation to one another. Interestingly, in extreme forms they can be almost identical.

For its part, the input span in an organization's control net or nets, when reduced, increases the density of forcible maps and hence the tractability of repetitive behavior. It does this at the cost of variety in structural options. However, where the informational load on an organization is high, thus perhaps constraining input span itself to be high, tractable system behavior can be obtained, on average, without modification of the control structure (the suggestion is) if relatively high intensities of MBE or of MBP, or relatively low levels of MBCL, can be provided.

What might the interrelationships between the three managerial control styles imply for organizational theory? As an example, we first observe that any given mapping necessarily has definite levels of internal homogeneity, forcibility, and threshold. Hence the implication can be drawn that any organizational unit with a fixed functional regime operates at definite levels of management by exception, management by priority, and management by consensus level, all at the same time.

To continue the illustration, let it now be assumed that these managerial styles each have a distinct psychological effect on the unit's work force. If functional regimes (mappings) should affect workers in such a manner, it then becomes an important theoretical point how free a manager might be to jointly vary the three control styles. The joint relation between the three control variables discussed in Sec. 4.6 shows that the three strategies are correlated, but much less than perfectly. Therefore, it should be possible to manipulate the density of, let us say, forcible mappings, either directly or by acting to vary the intensity of management by priority, or indirectly by changing the intensity of either exception or priority management, while at the same time allowing for fine-tuning of some other organizational response (e.g., organizational climate) by modifying the intensity of a remaining strategy.

To illustrate the last point, recall that Kauffman suggested 60% as a density of forcible mappings which ensures that a net's repetitive behaviors will typically be tractable (see Chapter 5). Although we should not take that particular figure as certain, it not being clear that Kauffman meant it to be so construed, let us use it to illustrate the combined use of strategies. Assume that an organizational control net has an input span of 4, and that an intervention to decrease consensus level has achieved a threshold level of 3 in the net. If no information is available as to the existing levels of internal homogeneity and forcibility, a reasonable estimate of the density of forcible maps is that provided by the marginal distribution of forcibility given a threshold of 3. Referring to Table 4.5, which shows the joint relation between the variables, the predicted density of forcible maps is 20%: lower than the desired 60%. If a second intervention can raise the intensity of exception management so that internal homogeneity is at least 13, while not raising the consensus level, the organization is assured of at least 77% forcible maps in the control net. (More correctly, the organization is assured that 77% of the maps in the population from which the net maps are assumed to be a random sample have at least one forcible input. For large nets the population figure is a reasonable prediction as to what will prevail in a given net.) That organization's particular control system is now constrained in the sense that usable repetitive behaviors are a practical possibility. If desired, management by priority can still be used to modify the organizational climate, or to accomplish other aims.

Naturally, these theoretical comments raise questions. Do real organizations suffer pathologies which can be interpreted as increased cycle lengths in a behavior space? Do our theoretical control regimes really have distinctive and consistent psychological effects in control units? If so, what are they? Before we attempt to suggest any answers it would be useful to sketch in an informal way how the theoretical position developed so far might work in practice.

In Practice. Consider the following hypothetical conversation. A consultant (C) has been called in by the general manager (GM) of a large division of a corporation.

GM: Our problem is that we are having increasing trouble meeting our deadlines. It's taking us longer than it should to execute routine.

C: Are there any assignable causes, such as inventory shortages, labor problems, or communication breakdowns? Any specific difficulties?

GM: I've been checking on that kind of thing for two weeks now. I can't find any definite reason for what is happening. I'm beginning to think that something more general is at fault.

C: How much time do we have?

GM: Not much. What can you suggest that can be done in one day?

Comment: C suspects at this point that "engineering" a solution, that is, debugging the organization's operations to determine specific remedial interventions, even if limited to any GM might have overlooked, would take more time than is available. C also agrees with GM's thought that a localized remedy may not exist at all. Whatever should be done, it now appears, will have to be accomplished in the face of considerable uncertainty on both C's and GM's parts as to details of the organization's structure and the functions of its basic parts. C tentatively diagnoses the difficulty as excessive cycle length in the organization's repetitive behavior, produced by a widespread change in some influential parameter (or parameters) related to cycle length. C now seeks to determine which of these parameters is accessible to influence in the present setting.

C: Would it be feasible to ask your people to cut down the number of sources of information they use in executing routine? We would want them to cut that number down to three if they could.

GM: You want everybody to do that?

C: Ideally, yes.

GM: That sounds too hazardous to me. We need a great deal of coordination around here. I think we might unravel completely if we did that. Besides, the front office doesn't like radical changes in organization.

Comment: C has suggested managing input span. GM decisively rejects the suggestion. C elects to focus on management strategies which emphasize function rather than structure.

C: Could we ask your people generally to decrease the amount of information they require to take action--or consistent inaction, for that matter?

GM: Won't that mean they'll be operating on the basis of less total information? That just doesn't *sound* right to me.

C: But you will agree, won't you, that *something* has to change? Perhaps we can just suggest that they try to be more decisive.

GM: Well, maybe. The front office does like to encourage a bit of hard-driving attitude. What will that accomplish here?

C: We can't be absolutely sure that will speed things up for you. But we do know from indirect theoretical evidence that there's a chance that this kind of change will do it.

GM: There's a problem here, if I understand what you're saying. Some of my people may not feel comfortable just adding up information sources or taking increased risk in that regard. These people are my technical types. It may seem inappropriate for them to consider their various information sources as interchangeable, or to act, in their minds, rashly. Is there anything else we could do?

Comment: C has suggested management by consensus level, first rather formally, then by reference to a psychological thrust possibly associated with it. GM accepts the suggestion only partially. C now offers two alternative approaches.

C: You do understand what I have in mind. We could suggest that they be somewhat more trusting in the way they make decisions in executing routine, perhaps by delegating tasks, or that they increase the number of people they take more seriously.

GM: Knowing these people as I do, I like the first one more than the second. But it sounds like the same thing you said before. These are all ways of taking additional risk, or increasing the possibility of making a mistake, or something like that.

C: We're not really sure *what* they are. We know they aren't *exactly* the same. But they do have certain things in common. Their overall effects are a lot alike, for one thing.

GM: I suppose we could do it. It seems awfully inexact. But I may not have much choice.

Comment: C has suggested management by exception and priority, respectively, phrasing them in terms of possible psychological thrusts for the sake of communication. In so doing, C has, of course, gone far out in front of firm justification. GM clearly prefers an engineered solution, but reluctantly agrees to the probabilistic tactics offered by C.

C: O.K. What we'll do now is aim toward putting changes of the sort you think most acceptable in place as widely as possible. Can we get your people together this afternoon for a meeting?

Speculative Interpretation. The reader should regard this section separately. In the interpretation above, we attempted to keep our work disciplined by accepted knowledge. Here we relax somewhat and seek to discover how far this modeling approach might help extend theory, and what the character of that theory might be, so extended.

The discussion of the joint use of the three strategies used the fact that MBE, MBP, and MBCL are imperfectly correlated. The joint correlation, however, is fairly high, and this (possibly) has its own implication for organizational psychology. Whatever the distinct psychological effect of each strategy might be, routine-oriented functioning organizations which are also small-scale mechanisms (see Sec. 6.6) would appear to offer their control system workers only a limited subset of all possible psychological environments. To escape these psychological limitations, either large-scale structural or functional arrangements would be required (i.e., the organization recast in large-scale) or more direct manipulation of the motivational climate in the organization undertaken.

What might the psychological character of each strategy be? In passing, note that the number of inputs, being related to the potential volume of information affecting the control unit, may well have its own effect on psychological climate.

To develop an interpretation for extended threshold, we focus on the control unit's input as supplying information sufficient for some consistent commitment of that unit. As threshold is lowered, the amount of potentially relevant information used to sustain the same commitment decreases. It would appear then that a decreased threshold implies the acceptance, somewhere in the unit, of greater risk. Hence MBCL might be associated with a timidity-boldness or a risk-taking continuum.

When forcibility exists in a control unit's mapping, the unit's output is linked decisively to certain unit inputs. Since in a control regime of that sort one might look most often to the forcible inputs for direction, units with greater forcibility might show more status-related behavior or more elaborate status structures. Hence high MBP will be assumed to produce, or require, higher degrees of status consciousness or other status-related behavior.

Internal homogeneity, when high in a control unit's mapping, suggests an attempt to control with a certain steady, noninterventionist, "what we usually do is O.K." air. Hence high MBE might produce, or require, a more trusting climate in a control unit so governed. Low levels of MBE, as interpreted here, might be characterized by a higher activity level, a more frantic, work-obsessed climate.

Under these interpretations, the conventionally productive complex organization with high input span could theoretically appear in two forms, depending on whether it were a "small-scale" mechanism or not. (1) If not, its control nets would be specifically built in large-scale detail to provide the required tractability of its productive routine. Such organizations' control units would be able to show a wide variety of psychological environments, since their control regimes would not be constrained to the "desirable" mapping types of constrained networks. However, in these large-scale networks, we would expect net structure itself to be importantly controlled. That is, such organizations would show the presumably few large-scale organizational forms both socially congenial and supportive of tractable system behavior. Alternatively, these organizations would be found to invest organizational resources relatively heavily in the active maintenance of less felicitous organizational forms which still control behavior suitably.

Or (2) the organization might be a small-scale mechanism. In this case, we would expect to see more variety in structural form, within the organization as a whole, perhaps, but certainly between the given organizational system and other similar examples of small-scale network mechanisms. We would also expect the organization, overall, to invest less in procedures for and interest in the maintenance of structural forms. (Within the organization, structure might well still be a matter of interest, but seen as more of an issue of local concern.) At the same time, we would expect less variety in control units' psychological environments. The environments expected in these control units would be such as to suit personnel who can function well at rather high levels of trust, who prefer at least a modest amount of status structure in their workplace, and who can tolerate moderate to high levels of risk.

The observations above apply to the "conventionally productive" organization, one that requires its work routines to have a relatively short cycle length. Are there organizations which might want at least some control systems to have very long cycles or run-ins? Perhaps so; particularly in control systems whose output is or requires the use of novelty or originality. As examples, consider product development laboratories, advertising copy groups, research departments (depending on the type of research, of course), and cooperative creative activities generally. In these settings the small-scale network would be expected to have a high input span, to bother very little with organizational form maintenance, and to have control units in which few work-related status structures are found and control unit people who are more work obsessed and less willing to take risks than are workers in conventionally productive units.

It is tempting to extend this theorizing into the sociopolitical domain and ask how the cycle length of human routine varies with the character, size, and input span of the interaction group; how much variety is shown, and how much societal investment is made in interaction structures; how routine and creative social systems differ with respect to static characteristics; and so forth. However, for us to make the extension would require so much speculation as to weaken the case we are trying to make for the utility of the general modeling technique. Therefore, we leave the topic with the hope that it will be pursued by students more familiar with the subject than are we.

6.8 An Advertising or Marketing Model

In much the same spirit as guided Sec. 6.7, we wish to illustrate in this section how small-scale, network based modeling methods might be used in marketing. Using the context developed in Sec. 6.6.4, we suggest a model for advertising influence in groups of people whose interactions mutually determine their attitudes. Assuming the interaction structure to be unknown in detail, but fixed, we take the familiar switching net model as the generic form, using ensemble-based modeling to handle the unknowns of net structure and net element function.

In particular, our focus is on opinions, favorable or unfavorable, held by home-buyer influence groups with respect to a specific real estate broker or brokerage firm. The model assumes that within a given home-buyer group the influence structure is (relatively) fixed and that some advertising efforts can be considered to be directed at how individuals in an influence group interpret word-of-mouth opinion, rather than being directed at modifying the group's existing transaction pattern.

The foregoing leads us to suggest that the reactions of an individual home buyer to others' opinions of the broker may be modeled via a Boolean function. In this representation, function input lines represent links to those that influence the home buyer, and function output lines represent links to those the home buyer influences.

The interactive nature of the entire home-buyer group is modeled by a set of appropriately chosen functions connected together in a network. As a first theoretical step, the structural ensembles used are of the familiar, flat, bench-mark variety, with the following qualification. In this model particularly it appears to make sense to insist that the sources influencing any given buyer shall all be different, that is, no one source can have more than one "effect" on the same buyer. The structural ensembles used for the networks therefore are generated by drawing samples for the k inputs to each net element without replacement. This conforms to the theoretical development in Sec. 4.7.1.

The group's collective opinion over time, which the broker wishes to influence, is then represented by the time course of the network's behavior. Since the broker is interested in the collective opinion of the cohort, he or she is less interested in whether or not group behavior is cyclic, and more interested in how many favorable dispositions exist in the group at any given time.

Our 0-1 interpretation is asymmetric: 1 represents a favorable attitude toward the broker, while 0 can represent either a neutral or an unfavorable disposition. Clearly, 0's and 1's are not only different, but differ in outcome significance as well. Outcomes are to be measured by how many 1's exist in the net at a given time.

It should be clear that the model's interpretation can be extended more broadly to dispositions toward, for example, a specific product or brand name. It is in this sense that the distance between net states can itself be interpreted as a measure of "brand loyalty." States close to one another indicate a higher degree of brand loyalty in that few people have changed their attitudes toward the brand (or broker) as the group moved from the one collective attitude to the other.*

---

*In some contexts brand loyalty might be better referenced by a measure of "1"-to-"1" transitions only.

One global restriction on the interpretation of the model lies in its fitting either a noncompetitive environment (one broker, for example) or a market in which for whatever reason the buyers' responses to the advertising of the one broker are quite independent of competitive activities.

The initial-state ensembles used in the simulations and the theoretical development of this model fit a "new broker, known initial position" scenario. It is assumed that the buyer group has had no history of interaction with respect to the particular broker (or brand). Thus, with this model, it makes modeling sense to start the model net in the behavioral space at points that are in some sense arbitrary; these initial-state ensembles are not technical ensembles.

We also assume that the broker knows his or her preference level before the advertising campaign starts. This typically would be established by a market survey. In the simulation reported in Sec. 5.3.4 we examined the effects of low, medium, and high beginning acceptance levels, using flat ensembles of initial states with 25%, 50%, and 75% 1's, respectively.

In our simulation we looked at buyer interaction groups of three sizes: 12, 36, and 108; and input spans of 1, 2, 3, 5, and 7. We hoped thereby to bracket realistic values, and to allow the examination of behavioral effects parametrically.

To summarize the simulation, we made 100 independent replications of "word-of-mouth"-produced response to "advertising" in each type of buyer group. That is, we examined the typical responses of buyer groups of a given size, initial preference level, and input span, recording both the time course of collective disposition toward the broker (or brand) and "brand loyalty."

Note that the data we thus obtained do *not* directly apply to the response of the broker's market area as a whole. The typical market area is probably composed of several, perhaps many, buyer groups of various sizes, in each of which input spans may differ, and so forth. Indeed, even the initial market survey results do not give the initial preference levels of each of the different buyer groups exactly. To predict the outcome in the broker's market area would require that more be known about either his particular market, or the ensemble characteristic applying to his market if the details are unavailable. Our data can be used as derived to get some idea as to the efficacy of different advertising approaches (as interpreted here), to provide theoretical guidance as to what buyer group characteristics are influential, and to help deduce what should happen in more realistic markets.

The Formal Controls. The formal functional controls of interest in this section make use of the asymmetric versions of internal homogeneity, forcibility, and extended threshold introduced in Chapter 4. We will concern ourselves exclusively here with these asymmetric control schemes. For that reason, we will not bother to distinguish them in our terminology from the symmetric forms. To review, the (asymmetric) internal homogeneity of a mapping is simply the number of 1's present among the output possibilities of the mapping. Its forcibility is the number of inputs in which a 1 being present in any of them forces the mapping to a 1 regardless of the circumstances at other inputs. Its threshold is the (smallest) number of "incoming" 1's required to ensure a 1 output from the mapping; that is, the output is 1 if that number, or more, 1's are present.

The Formal Controls as Types of Advertising Message Content. The preceding paragraph has reviewed the basic definition of the measures on which our "advertising" controls are built. We will now particularize these measures in ways that provide more realistic interpretation as types of advertising content.

It appears unrealistic to think that any kind of advertising could have the effect of ensuring a *specific* level of, say, internal homogeneity in all net mappings. We will assume that the advertising copywriter can guarantee at least (or at most) a certain level of internal homogeneity, forcibility, or threshold in any given net mapping. Other values of the measures are determined in some interpretable probabilistic manner. That is, we specify three kinds of functional ensembles, each building on one of the three mapping measures, and each interpreted as a type of message content.

Primary Persuasion. Our advertising control which influences via internal homogeneity consists of functional ensembles in which individual mappings are chosen as follows. For example, if the level of control is 5, there will be at least five 1's in the mapping. If in the example, $k = 3$, since the mappings will have eight rows, there *could* be 5, 6, 7, or 8 1's in the mapping. In determining the given mapping, a choice is first made as to this point. In our scheme, 5 would have the greatest probability of being chosen, 6 exactly half that, 7 exactly half of *that*, and so on. That is, we are thinking of an ad copywriter who can guarantee internal homogeneity of at least 5, and for whom actually achieving higher values than 5 is increasingly less likely. The second step in determining a given mapping is just to permute the chosen number of 1's (and the appropriate number of 0's) in the mapping at random.

Since it is very unlikely that any ad copywriter designs a message with the intent of increasing internal homogeneity as such, it is natural to ask what an approach to message formulation that approximates the mapping assignment scheme just described might be like. Recall that whatever the particular strength of the message effect *in the large*, it works without the existence of any systematic relationship between output and input. It seems therefore to accomplish what opinion molders might often want to increase, namely general reactivity to something, without much consideration of the settings in which those reactions occur. Hence our name for it: primary persuasion.

We do not claim expertise in this area, but it appears to us that substantial amounts of simple "exposure" advertising might be of this genre.

Imitative Persuasion. Our advertising control ensemble which acts through forcibility can be described rather easily. Each mapping is determined as follows. A set of inputs of size equal to the control level is chosen at random (without replacement) from the k mapping inputs. The mapping rows are set at 1, in which the chosen inputs have 1's. The remaining rows are set to 0 or 1 equiprobably. We are here thinking of an ad writer who can guarantee that if a "favorable" opinion is held by anyone of the chosen set of information sources, the buyer produces a "favorable" response. In this scheme each of the chosen sources is a "forcing" input, with 1's forcing 1's. Forcing inputs in addition to those chosen explicitly occur probabilistically.

How should we interpret this control style? This control style decisively links the "favorable" opinions of a set of influence sources to a similar response on the part of the buyer. That is, the control level sets the (minimum) number of "others" who are significant to the buyer with respect to the broker or product in question. The end result is that, as far as positive reaction to the broker or brand is concerned, the buyer imitates these significant others. Messages pointed toward effecting this reactivity in buyers would persuade them of the value of that sort of imitation. Hence our label: imitative persuasion.

We speculate that this kind of control might be produced by advertising in which generic character types are shown approving the product-directed activities of those to whom they are likely to be significant. The telephone company ads in which family members respond enthusiastically to the long distance calls placed by their children, siblings, or parents might be examples. That the ads also point out how lower rates may be obtained could be interpreted as an effort to get an additional "primary persuasive" component into the message as well.

Consensus Persuasion. This control is even more straightforward than the previous two. In specifying a given mapping, the control level gives the maximum number of inputs showing a "favorable" opinion required to produce a "favorable" response in the buyer. In the given mapping, all those rows with the control level, or more, 1's among the k inputs, are set to have output 1. Other rows are set to 0 or 1 equiprobably. Thus it is possible that a given mapping might show "favorable" response to consistently *fewer* than the control level number of "favorable" opinions. Here we are thinking of an ad writer who can guarantee a favorable response in buyers if a *certain number* of *any* of their influence sources are favorable to the product. The ability to produce the same response consistently to fewer than that number of favorable opinions is a matter of luck.

We call this message content, consensus persuasion. Messages with this thrust would be recognizable in their attempt to produce in buyers a sensitivity to quantity of opinion; to promote a potentiality to ride the proverbial bandwagon. The control level gives the number of passengers carried sufficient to make any buyer climb aboard. For example (the control level is 2), "if two or more of my influence sources are in favor, so am I."

We suggest that ads with a convivial air might be messages with this "consensus persuasive" content. An example of a gemütlichkeit-oriented message is the television beer ad which features the happy conclusion: "We're in Schaefer City." The special-occasion-means-Brand-X messages (e.g., "Tonight let it be Lowenbrau"), on the other hand, seem more likely to be examples of imitative persuasion. As pointed out above, we do not mean to suggest that messages can carry only one type of content.

Buyer Group Dynamics. Having sketched the static model and ensemble linkages, let us turn to the simulation results and examine them in context. These results apply to functional ensembles defined in the particular ways we have defined them. We have used relatively interpretable, but in some respects arbitrary control schemes which make use of the measures internal homogeneity, forcibility, and threshold. Quite aside from the possibility that different measures exist, it should be remembered that other ways of defining controls using those measures are also possible. These results have been discussed in Chapters 4 and 5. Here we will emphasize what appear to be the more noteworthy suggestions for marketing theory and practice.

Initial Market Position. The results indicate that the broker's (or the product's) level of initial acceptance can affect both mean acceptance and mean brand loyalty. This effect largely washes out in the long term. In the short term the broker's acceptance level is, not surprisingly, higher with higher initial acceptance. (Note that this short-term effect does not apply to primary persuasion, under which acceptance immediately reaches its terminal level.)

Input Span. Although input span (k) can affect terminal acceptance levels, its most dramatic effect is on brand loyalty. In summary, the broker or product-line manager achieves higher brand loyalty in buyer groups with a smaller input span.

This raises the question as to whether a manager might want to consider advertising as a means to manipulate the input span directly. We suggest that this course of action is not only theoretically plausible, but that it may even be observable in practice. Consider, for example, the Camel cigarette ads which feature, amid the admiring glances of attractive young women, vigorous men who *do* smoke Camels, but who emphatically do *not* "follow the crowd." What we may see in this message is an attempt both to increase the level of imitative persuasion and to reduce the input span to 1. Indeed, this message, in addition, may be attempting a more radical structural modification of buyer groups: not only is the input span to be set to 1, but the buyer's sole influence source is to be none other than the buyer himself. We do not need to simulate such a buyer group. Modeled as a net, it is an extreme structural and functional form in which brand loyalty levels immediately reach and retain the maximum level.

Effects of Message Type. Considering only those levels of control examined in our simulations, imitative and consensus persuasion appear more powerful than primary persuasion in moving both acceptance and brand loyalty levels toward desirable outcomes. Some levels of consensus persuasion are notably ineffective, especially k, and to some extent k - 1. Otherwise, consensus persuasion is a strong control. Imitative persuasion appears to be a very strong control, so strong that on looking at preliminary results we felt obliged to simulate only a single control level. What this suggests for imitative persuasion is that a more realistic definition of this variable would allow control levels even lower.

Clearly, we run into difficulty in attempting directly to contrast message types. To do this in general requires some rational way of equating the various levels of the different persuasive styles. One approach would be to assess the likelihood in practice of achieving the various levels and to equate those levels which are equally likely. We can approximate that method to a limited extent here by observing that in practice it might be nearly impossible to achieve advertising messages whose impact would be profound enough to yield *very* high levels of control. The simulation data suggest that our imitative persuasion of level 1 is already a high and possibly very high level of control. On the other hand, at least some of our lower levels of control are plausibly attainable.

Fortunately, perhaps the clearest difference in effect among the message types is broad enough to avoid these scaling difficulties. Primary persuasion achieves its terminal level of approval, whatever that level may be, immediately. Both imitative and consensus persuasion, on the other hand, except at high intensities, achieve their final levels of approval, which may be quite high, more or less gradually.

This difference is important. If circumstances are right, primary persuasion has a decisive advantage over the other persuasive styles. Using messages capable of producing a fairly high level of primary persuasion but only moderate levels of imitative or consensus persuasion, if the advertising budget or other considerations requires the ad campaign to be short, primary persuasion is the clear choice.

Among the other considerations relevant in the design of an advertising campaign would be the relative persistence of the effects of the different message types. Our results do not speak to this point, of course, but we find it impossible to resist the speculation that, of the three, primary persuasion would be the style *least* resistant to extinction, precisely because in primary persuasion output events are not systematically linked to input events. Because of this, output events are also less likely to be linked in individual buyers' associative structures to broad features attending those events. Conversely, consensus, and perhaps especially imitative persuasion, *do* link output events into input events systematically and hence broadly. As a result, we might expect the functional forms induced by these message types to be more persistent.

We might therefore suggest to our hypothetical broker that, other things being equal, if he or she wants primarily a quick but transient sales boost (assuming that sales will follow from a rise in word-of-mouth produced approval) then a short campaign making use of high intensity primary persuasion messages is indicated. On the other hand, if the broker is interested in relatively enduring enhancement of, say, the corporate image, consideration should be given to the feasibility of a longer campaign using consensus or imitative persuasion.

References

Dertouzos, M. L. (1965). *Threshold Logic: A Synthesis Approach*. Cambridge, Mass.: MIT Press.

Gelfand, A. E. and Walker, C. C. (1980). A system theoretic approach to the management of complex organizations: Management by consensus level and its interaction with other management strategies. *Behavioral Science*, *25*, 250-260.

Hempel, D. J. (1969). The role of the real estate broker in the home buying process. CREUES Real Estate Report 7, Center for Real Estate and Urban Economic Studies, University of Connecticut, Storrs.

Hempel, D. J. (1970). A comparative study of the home buying process in two Connecticut housing markets. CREUES Real Estate Report 10, Center for Real Estate and Urban Economic Studies, University of Connecticut.

Kauffman, S. A. (1969). Metabolic stability and epigenesis in randomly constructed genetic nets. *Journal of Theoretical Biology*, *22*, 437-467.

Kauffman, S. A. (1970). The organization of cellular genetic control systems. *Mathematics in the Life Sciences*, *3*, 63-116.

Kauffman, S. A. (1974). The large-scale structure and dynamics of gene control circuits: An ensemble approach. *Journal of Theoretical Biology*, *44*, 167-190.

Koontz, H. and O'Donnell, C. (1968). *Principles of Management* (4th ed.). New York: McGraw-Hill.

Taylor, F. W. (1947). Shop management. In *Scientific Management*. New York: Harper & Row.

Walker, C. C. and Gelfand, A. E. (1979). A system theoretic approach to the management of complex organizations: Management by exception, priority, and input span in a class of fixed-structure models. *Behavioral Science*, *24*, 112-120.

Index